# SOCIAL STUDIES OF SCIENCE

# SOCIAL STUDIES OF SCIENCE

**Bernard Barber**

**Transaction Publishers**
New Brunswick (U.S.A.) and London (U.K.)

Library of Congress Catalog Number: 89-20590
ISBN: 0-88738-329-7
Printed in the United States of America

**Library of Congress Cataloging-in-Publication Data**
Barber, Bernard.
    Social studies of science / Bernard Barber.
        p.    cm.
    Includes bibliographical references.
    ISBN 0-88738-329-7
    1.  Science—Social aspects.    I.  Title.
Q175.55.B371990
303.48′3—dc20                                  89-20590
                                                         CIP

For Jeffrey C. Alexander, Jonathan R. Cole, Daniel Sullivan,

and Viviana A. R. Zelizer

Cherished Friends and Colleagues

# Contents

# Introduction

## Multiple, Diverse, and Unexpected Origins: Toward an Analytical Sociology of the Sociology of Science

One of the essential tasks of the sociology of science is to develop an empirically based theory of that perennial phenomenon in science, the emergence and development of new scientific specialties. We now know that science is not a uniform, homogeneous body of knowledge but a great and complex network of partially independent but also interdependent specialties. It is only when viewed from afar that science takes on its unitary form. Close up, we see many and diverse clusters and groups of scientists working away on their specialties, building on special accumulations of theory and data, seeking innovation through special images and methodologies. Science as a whole moves through the movement of these specialties.

The sociology of science is one such specialty in that cluster of specialties known as the discipline of sociology, which itself has its place in the larger area known as the social sciences, which in turn have for some hundred years or more aspired to be a part of that largest whole of all, unitary science. Paradoxical though the enterprise may seem at first glance, if we use the emergence and development of the sociology of science itself as a case history for building a theory of the general processes of the emergence and development of scientific specialties, we may be able to contribute to that theory.

It has been my good fortune as a sociologist to have been connected with the sociology of science in a variety of ways for more than fifty years. Indeed, I can be said to have been present, if only in a minimal way, at an important part of its very creation, before it was even thought of as a specialty. Also, a little later, I was an early participant, through my book *Science and the Social Order,*[1] in the effort to bring together the enormously diverse knowledge that

1

might constitute the basis for a self-aware specialty. And, finally, the essays collected in this volume bear testimony, I hope, to my continuing though intermittent efforts to contribute to a specialty that has now flowered and become very much aware of itself. The sociology of science today has its clusters and groups of full-time specialists, its own professional meetings, journals, prizes, and the other typical products of a scientific specialty, such as diverse theories, methods, and competition among schools. From scattered but distinguished beginnings, it has arrived as a mature specialty. We study this creation, development, and arrival to see what we can learn of the general processes of specialty formation.

It was quite by chance that I was present at an important part of its creation and am able therefore to recount something (but far from everything, of course; I hope this essay will stimulate further accounts) of how I think the sociology of science came to emerge as a specialty. As an undergraduate student in Harvard College from 1935 to 1939, I happened to encounter, as teachers, scholars who had (each for quite different reasons) an interest that resulted in writing on the social aspects of science. There was Pitirim A. Sorokin, from whom I took introductory sociology; there was my tutor in sociology, Robert K. Merton; there was Talcott Parsons, from whom I took several courses, all of them concerned in one way or another with the problems of rationality and science in social systems; and, finally, there was Arthur M. Schlesinger, Sr., from whom I took a course in the social and intellectual history of the United States. It was in that course that I wrote my first research paper in the sociology of science, on Thomas Jefferson as a scientist.

In the next section of this volume—part I: Historical Origins and Development of the Sociology of Science—the emphasis in the three essays included there is more a historical one, though some analysis is inevitably included. Sociologists cannot help doing their thing! In this introduction, the purpose is *entirely analytical* and the historical materials, which overlap somewhat with those in part I, are intended to give some validation to the analytic goal. As I have already said, this introduction is intended to be one case history in the effort to construct a general theory of the processes of the emergence and development of scientific specialties. Since such emergence and development occurs through a variety of processes (for one process not mentioned in this introduction, for example, see the essay in part II on fashion in science), we shall need many analytical case histories to develop the general theory I have in mind. And it should come as no surprise if such a theory has something to say to a still more general theory, that of the nature of change processes in social structure and culture in general.

Before proceeding to state the general analytical propositions that are supported by the data presented in this introduction, another word of obvious but necessary caution. This is not the complete and intensive account or accounts that may someday emerge from the writings of sociological historians of the sociology of science. Nearly all sociology, not just the sociology of science, still lacks that kind of account. This is to be construed more as a sketch of what such an account might look like, a contribution by one participant observer to such an account. While it has obvious limitations, inevitably, even such a single person's participant observation has advantages that are lacking to the historian who comes later and relies entirely on archival sources. Both the report of this participant observer and the archival materials, plus such other oral histories as may be available from other participant observers, are what is really needed for a wholly successful analytical account.

What, then, are the several analytical propositions about the emergence and development of a scientific specialty that are illustrated by my account, which is based on my experience as a long-time participant observer, of the sociology of science? They can be stated succinctly as follows:

1. There is no single source, for example, "the inherent dynamic of accumulating knowledge," for the sociology of science. Quite the contrary. It originated, developed, and experienced acceptance and resistance as a result of multiple, diverse, and sometimes unexpected social-structural and cultural (the personality factor is here omitted) factors and processes in the environing social system. The economy, the polity, the educational system, values, ideologies, and of course sociological knowledge and research themselves all played their interacting parts.

2. The proposition above about multiple, diverse, and unexpected origins holds not only for the work of the aggregate of contributors to the sociology of science but also to any particular scholar. Each scholar sometimes had multiple, diverse, and unexpected determinants of his or her work.

3. In the very early stage of the emergence and development of the sociology of science, those who are later seen as contributors, even "founding fathers," may have been unaware of the relation of their work to an emerging specialty; they may not have intended to establish a new specialty but, rather, to have worked in an existing one.

4. At different stages of the movement of the sociology of science from relative immaturity to relative maturity, different and differently combined determinants were operating. Certainly, awareness and deliberate intention increased. Formal educational mechanisms, with specialized groups and institutes devoted to the sociology of science, increased in relative importance.

Now, our actual account:

There is seldom an absolute beginning or act of creation for a scientific specialty. Forerunners, antecedents, foreshadowings can almost always be found, especially if someone wishes to do so. So antecedents of the sociology of science can surely be found before the 1930s, when my account begins.

But because several important events did occur then, I shall start there. It was in the 1930s that a massive critique was first made of the value of "pure science" and its associated ideology, a value and ideology that were widely and strongly held among working scientists at that time. This value and belief probably came to predominate in the late nineteenth and early twentieth centuries when basic science was trying to gain a measure of autonomy for itself, a measure that we now very much take for granted and that therefore does not have to be so absolutely defended.

The critique came primarily from a group of eminent British scientists, at least two of whom had won Nobel Prizes in physics, who came to be known as the "scientific humanists," men like J. D. Bernal,[2] Frederick Soddy,[3] and Lancelot Hogben.[4] The scientific humanists were interested in both sides of the social aspects of science: first, the effect of society on science and, second, "the social function of science," as Bernal called it, or the effects and uses of science in society. Bernal's book of that title is the classic, most complete, and most powerfully argued statement; a sampling of a few of the titles of the chapters of his book and their contents tell us the main views and concerns of all these challengers of the established "myth" of pure science.

In his introductory chapter, Bernal discusses the "interaction of science and society," "science as power," and "the scientist as worker," all very new views to the science of his day. The succeeding chapter is "historical," describing the long history of the interaction of science and society. Later chapters take as their subjects "Science in Education," "The Application of Science," "Science and War," "What Science Could Do," "Science in the Service of Man," and "Science and Social Transformation."

The scientific humanists were horrified by the social disaster of the 1930s' depression and by what they thought was also an accompanying general "retreat from reason" and from its chief embodiment, science. They saw science as a guide to and a means for social reform. They were influenced not only by Marxist theory but also by what they thought was the large and beneficial use of science in Marxist Russia at that time. They admired how the Russians were planning science itself and also using it in planning social development in general, as well as in selected programs in industry and agriculture.

Marxism had its most direct impact on them through the papers read by the Russian delegation to the Second International Congress on the History of Science that was held in London in 1931. The reading of these papers had a shock effect on their auditors for their explicit assumptions about the interaction of science and society. Most impressive of all the papers, apparently, was Boris Hessen's "The Social Roots of Newton's 'Principia.'"[5] For the scientists of the day, and for some time thereafter, Newton's *Principia* was the divine elixir of pure science. To argue its social origins and social consequences, as Hessen did so well, was a most striking indication of the need for a socially involved and socially useful science. The scientific humanists were inspired by Hessen; and their own subsequent writings, especially those of Bernal, came to have a large influence on other natural scientists and on social scientists as well. Thus political values and interests played an important role in this early and still unselfconscious phase of the sociology of science as a specialty.

The challenge of the scientific humanists to the absolutization of pure science did not itself go unchallenged. Pure science is supported by a strong set of values, and those who held them were offended by Bernal and his associates. They were especially offended by Bernal's seemingly simplistic conviction as to the possibilities for planning the development of science. A careful reading of Bernal's *The Social Function of Science*, however, shows he was well aware of the need for pure as well as applied science, wanting only to alter the existing balance between them in the direction of more planned and applied science. But his opponents did not look for subtleties of analysis, so outraged were they at the mere thought of any planning at all.

As a result of their outrage, in 1940 another group of distinguished British scientists, led by the brilliant Michael Polanyi, then a physicist but later to become a knighted philosopher, a social philosopher, and quasi-sociologist of science,[6] founded the Society for Freedom in Science in specific opposition to the supposed extreme views on planning of Bernal and his fellow scientific humanists. By June 1946 the society had 450 members—250 in Britain, 176 in the United States (where the Harvard Nobel physicist, Percy Bridgman, was the chief figure), and the rest scattered around the world.[7] Absolutized views of pure science did not yield easily, but out of this ideological conflict eventually came a somewhat better sociological understanding of the possibilities and limitations for predicting and planning science, both its discoveries and its social consequences. Contrary to certain idealistic views of the peaceful progress of science, conflicts may sometimes have useful functions by sharpening up the formulation of essential issues.

The ideology of pure science lived on in other quarters than those of the Society for Freedom in Science. Although there was very little academic history of science in either Britain or the United States in the 1930s, this scholarly specialty began to flourish and expand greatly after World War II. Until the late 1960s most members of this group practiced what came to be called, in the polemics of the time, "internalist" history, which was construed to be in explicit opposition to something called "externalist" history. Internalist history of science saw scientific ideas as developing and accumulating only through their own internal dynamics, with "great men" or "geniuses" being the chief instruments of these internal dynamics. Work that emphasized or included the social determinants of scientific advance was criticized and rejected as externalist. The internalist-externalist dichotomy and polemics, which now seems incredible to the sociology of science, was pretty much given up by its adherents in the late 1960s and early 1970s, though of course there is still some work that emphasizes the development of scientific ideas more than it does the social determinants of those ideas or the social organizations that sponsored and supported the men who created those ideas. The history of science now produces much excellent sociological history, all of which provides valuable data for an explicitly analytical sociology of science as well as serving its own purposes. It seems to me that a turning-point in this internalist-externalist conflict came with the publication in 1962 of Thomas Kuhn's *The Structure of Scientific Revolutions*,[8] of which more will be said later. It was not unimportant that Kuhn was himself a historian of science.

In addition to the scientific humanists and their opponents, there were some academic scholars in the 1930s who were important for what became the sociology of science. Chief among these were Pitirim A. Sorokin, Robert K. Merton, and Talcott Parsons, all of the Department of Sociology at Harvard, and James Bryant Conant, who in 1935 gave up his professorship of chemistry at Harvard to become its president. Since two essays in part I are full-length discussions of Sorokin's and Parsons's work, we can be brief in our treatment of them here, bringing out mainly our analytical interest in the development of the sociology of science. Merton and Conant we shall discuss at a little more length.

Pitirim A. Sorokin came to the United States in 1924 as a refugee from the Russian Revolution. Taking up a professorship at the University of Minnesota, he immediately established himself through his writings as an authority on rural sociology, then a major specialty at midwestern American universities. In addition, he published books on social mobility and the history of sociological theories. His breadth of scholarship, analytical talent, and remarkable productivity quickly put him in the very front rank of American sociologists. As a result, when Harvard University decided in 1931 to establish a Sociology Department, Sorokin was invited to become its first chairman.

At Harvard, Sorokin's interest turned to the large problem of the essential nature and typology of societal systems and their processes of development and change.[9] Using primarily the indicator of their basic epistemological commitment, Sorokin classified all societies into three types: "ideational," "idealistic," and "sensate." Each of these three types had its own characteristic dominant forms of *all* the following cultural and social structural subsystems (listed in the order in which Sorokin treated them in his four large volumes): painting, sculpture, architecture, music, literature, "systems of truth" (which included science), ethics, law, "basic categories of human thought" (which included causality, time, space, and number), social relationships, war, and revolution. A most inclusive conception of the substance and processes of social systems.

Volume 2 is where Sorokin dealt specifically with science. The "so-called 'scientific system of truth,'" said Sorokin, "is largely the truth of senses," which predominates in the sensate type of society. While science exists to some extent in all the three types of society, the "*crescendo, forte, and fortissimo* of the empirical system of truth" was from the sixteenth to the twentieth centuries. Sorokin's empirical evidence of the growth and changing predominance of science consisted of the data his assiduous scholarship had compiled on the "number and qualitative estimation of discoveries and inventions" from 800 B.C. to the present.

Thus Sorokin's demonstration of the social character of science came not out of any specialized interest in science but from his "grand theory" of the nature and development of societal systems as wholes. We should note, however, that unlike other grand theories of the time, Sorokin presented detailed and voluminous quantitative empirical evidence for his analysis. This use of quantitative data had, as we shall see, a strong influence on the work of his student Robert K. Merton.

Merton, who has been a central figure in the sociology of science from its emergence in unplanned and unselfconscious form in the 1930s to its highly mature condition as a sociological specialty fifty years later in the 1980s, came to Harvard in 1931 as one of Sorokin's first and eventually most distinguished students. Merton has been heard to say that he chose Harvard over other graduate sociology departments of the time because of Sorokin's book *Contemporary Sociological Theories*.[10]

Sorokin's influence on Merton was hardly at a distance. Sorokin hired Merton early on as a research assistant for his study of science; so valuable was Merton's help that Sorokin acknowledges it in two places. In a footnote to chapter 3 ("Movement of Scientific Discoveries and Technological Inventions") of his *Dynamics*, Sorokin says, "In co-operation with R. K. Merton and J. W. Boldyreff." In another footnote, to chapter 12 ("Fluctuation of General and Special Scientific Theories"), Sorokin says, "In co-operation with R. K. Merton." Furthermore, as a joint author with Sorokin, Merton published two papers coming out of his research for *Dynamics*: "The Course of Arabian Intellectual Development, 700–1300 A.D."[11] and "Social Time: A Methodological and Functional Analysis."[12] No wonder that, in the collection of his essays in the sociology of science, Merton dedicated the book "To my teachers: Pitirim A. Sorokin, Talcott Parsons, George Sarton, L. J. Henderson, and A. N. Whitehead."

*Science, Technology and Society in Seventeenth Century England*, Merton's 1936 thesis, published under the auspices of the founder of the history of science, George Sarton,[13] has been a landmark study ever since its publication. In 1988 a fiftieth-anniversary celebration in its honor was held in Jerusalem. It has called forth a whole literature of praise, criticism, and relevant scholarship.[14]

In his preface, Merton stated his purpose. In what is something of an overstatement, as our previous discussion has suggested, probably made out of scholarly modesty, Merton began by saying: "To be sure, the view that the science of any period is not divorced from its social and cultural context has become, properly enough, a commonplace. . . ." But then he went on to say that "there are few empirical studies of the relations which do obtain." In sum, his study "is an empirical examination of the genesis and development of some of the cultural values which underlie the large-scale pursuit of science." Or, more specifically, in another place he says, " . . . this study will be concerned with the sociological factors involved in the rise of modern science and technology."[15]

In his graduate student days, Merton had made himself a determined and already much admired academic scholar. His thesis was intended as a contribution to the science of sociology as a whole, not to an as yet unborn specialty, the sociology of science. Especially after this research with Sorokin, and surely with more general encouragement from the sociological culture of the period, he was eager to do an empirical study. But also, and again with influences from not only Sorokin but his other teacher, the theorist Talcott Parsons, he wanted to make a theoretical contribution. Although he eventually dealt with a variety of influences on seventeenth-century science—economic and political interests as well as religion and associated values and ideologies—Merton was especially concerned to show that, contrary to Marxian and other materialist theories of the time, religion and values had their degree of independent influence in social systems. As many scholars have viewed it since, it has its Weberian perspective.

Merton chose as his research site the remarkable development of science in seventeenth-century England. He did not take the fact of that development for granted. In a tour de force of single-handed scholarship, Merton classified the vocational interests of all 6,000-odd persons from the seventeenth century treated in the *Dictionary of National Biography* and *demonstrated* the increasing number of Britishers involved with the new activity of science. His tables, graphs, and charts detailing this rise of science were probably unique at the time for a work in sociological theory. In discursive language, he summed up his finding as follows: "From the middle of the seventeenth century, science and technology claimed an increasing need of attention. No longer an errant movement finding faltering expression in occasional discoveries, science had become accredited and organized."[16]

On the theoretical side, Merton went on to show that a disproportionate number of the new scientists were Puritan believers and that the Puritanism of the time consisted of a set of values and religious beliefs that were favorable to the practice of science. Thus Merton had extended Max Weber's argument about the influence of the Protestant ethic on capitalism, also constructed for general theoretical reasons, to the other important modern development, the great growth of science and technology. Furthermore, as a part of his theoretical analysis of the general processes of social change, Merton was at pains to point out: "One of the basic results of this study is the fact that the most significant influence of Puritanism on science was largely *unintended* by the Puritan leaders."[17] Thus, in a variety of ways, Merton's concern for an empirically based sociological theory, using the case of seventeenth-century English science, made lasting contributions to that theory and also provided an important prototype—not too often followed, unfortunately—for what became the sociology of science.

In the case of what have become two other landmarks in the field, his papers "Science and the Social Order"[18] and "Science and Technology in a Democratic Order,"[19] the social influences on Merton's 1930s work in the sociology of science were somewhat more complex, adding political and value/ideological interests to his interest in sociological theory for its own sake. In these two papers, which became seminal for the sociology of science, Merton was perhaps primarily concerned for the theoretical task of defining the central norms or values of science as a special kind of social activity. But he also expressed political and ideological horror, widespread among academic liberals of the 1930s, at the attacks on universalistic science by the Nazis and at their racist banishment of Jewish scientists. Merton discussed these threats in two sections of the first paper, "Sources of Hostility to Science" and "Social Pressures on Autonomy of Science." Merton's theoretical interest in the norms of science was heightened by his awareness of their fragility in the kind of society Nazi Germany had become.

So much for Merton's work in the 1930s and its diverse and multiple social origins. I shall return below to Merton's important later work and teaching in the sociology of science when it commences in the late 1950s.

As I said earlier, since an essay of mine on Talcott Parsons and the sociology of science is presented in part I, I can be brief here about the social sources of his work in this field. Parsons had a theoretical interest in rationality and science because of its importance for his attempt to develop a general theory of "action" and because of the central place of rationality and science in the modern social system. To understand rationality and science was to have a better understanding of the modern world and of the nature of social action in general. But Parsons also had a strong value/ideological interest. For him, rationality and science were essential for the kind of society he admired. He was a devoted citizen of the modern world as well as a sociologist. He saw no obstacle to being both a citizen and a scientist.

Parsons's position on citizenship and science was shared by James Bryant Conant, who came to the presidency of Harvard University in 1935 determined to democratize and nationalize it on universalistic and meritocratic principles. Among other programs to achieve his purposes, Conant immediately established the National Scholars Program to bring the best students to Cambridge, whether their parents could afford Harvard or not. But Conant's attention was soon drawn away from Harvard to the national problems arising in connection with World War II. An ardent supporter of the Allies, he was involved at the top levels of national science planning even before America's entrance into the war. And once America was in the war, he served at the top leadership levels in the development of the atomic bomb and other projects.

All of this activity made Conant more vividly aware of the importance of science and society for each other. Returning to Harvard, he defined it as his duty as a citizen and an activist educational leader to instruct, first, his Harvard students, and then the American public as a whole, in the importance of science to society.

His mode of instruction began with a set of case histories in the development of the various sciences, each case seeking to communicate what Conant, avoiding, as he said, the more abstract question, "What is science?" called "the tactics and strategy of science." In 1947, in the Terry Lectures at Yale University, intended for the general public, Conant described his purposes, explained how his cases had been developed and taught, and presented some of the cases already developed.[20] They were published as *On Understanding Science*. In the preface, Conant acknowledged the help of a group of young scientists and historians of science in developing the cases and in preparing and teaching his courses at Harvard. It should be noticed that one of these young scientists was the novice physicist, Thomas S. Kuhn, who, perhaps because of this experience, moved on first to the history of physical science and then to his important work in the philosophy and sociology of science. More about Kuhn's work in a moment.

One of the chapters in Conant's book was "The Scientific Education of the Layman." So successful, in fact, was Conant's book in educating the American laity, that a few years later he was asked to prepare a second edition. In response, Conant wrote a much enlarged book, *Science and Common Sense*,[21] this time expanding on his specific concern for the interrelations of science and society in two chapters: chapter 11, "The Impact of Science on Industry and Medicine" and chapter 12, "Science, Invention, and the State." If not only the American but even the world public now is concerned for the social aspects of science, if laypeople now have some sense of what has become the sociology and politics of science, we can attribute some of the origins of this concern to Conant's teaching and writing.[22] He did his part in democratizing and nationalizing an understanding of the social nature of science.

Since I have discussed at length in the first two essays of part I, which follows this introduction, the social and intellectual sources of my own initial major activity in the sociology of science—my 1952 book, *Science and the Social Order*—I can be brief about it here. For the analytical purpose of this introduction, I need say only that no visible, organized, self-identifying specialty in the sociology of science yet existed. I wrote my book as an exemplar of social-systems analysis, as an attempt to demostrate the intellectual and social-policy usefulness of the kind of empirically relevant social system theory I had learned from Talcott Parsons. I chose science as my exemplar not because it was especially valuable for this purpose; other areas of social structure and culture might have done as well. But, because of the happenstance of my familiarity with people like Sorokin, Parsons, Merton, the scientific humanists, and Conant, I was then most comfortable with material on this history and social aspects of science. Very few of the scores of books and articles I used and cited in my book referred to one another. There did not yet exist that dense and complex web of reciprocal and interwoven references and citations that indicates the presence of a scientific specialty. As I said in a review article on the sociology of science written only five years later:

> Despite its heritage and the steady stream of contributions from professional sociologists and others, the sociology of science is not now one of the most highly cultivated areas in sociology. . . . it has relatively few full-time research workers, and the volume of publication in the field is small. Until this year [1957], for example, no regular section of the annual meetings of the American Sociological Society has been devoted to the sociology of science. And to take one last indicator of the relative quietude of the field, there are few, if any, undergraduate or graduate courses devoted exclusively to the sociology of science.[23]

Just as I was writing this statement, however, and fortunately, the sociology of science as a specialty took some very large steps forward. One of the largest of these was the return of Robert K. Merton to intensive research in the field and, also very important, to graduate teaching of the subject. In his 1957 Presidential Address to the American Sociological Society, Merton signaled this return with his paper "Priorities in Scientific Discovery,"[24] which initiated a whole later career of work on the reward system of science: on the problems of multiple discoveries, patterns of evaluation, and processes of cumulative advantage (where Merton discussed what he called "the Matthew Effect," a now widely used idea).[25] This whole body of work and the associated work of Merton's very able students has been so influential that it has come to be described, eponymously, as "the Mertonian sociology of science."

In his graduate teaching at Columbia University, Merton was able to attract a set of original and productive young scholars, chief among them Harriet

Zuckerman and Jonathan and Stephen Cole. To Merton's primary use of historical materials in his own work, these younger scholars added the use of survey research (which had been developed so extensively at Columbia by Merton's colleague, Paul F. Lazarsfeld) to collect a variety of quantitative data on such topics as social stratification in science and the workings of the reward system as manifested in the Nobel Prize awards.[26] It should be noted that work by these and other Merton students was very much enriched by the invention in the 1960s, by Eugene Garfield, of the Science Citation Index.

Garfield was trained as an information scientist and invented the Science Citation Index as an aid to information retrieval for working scientists. Young sociologists of science like the Coles and others pioneered in using citation data as a useful, if not perfect, measure of the comparative scientific achievement and prestige of individuals, departments, and universities. Although he did not foresee this use of the Science Citation Index when he invented it, Garfield and the very successful organization he founded, the Institute of Scientific Information, in Philadelphia, have been most encouraging of this unexpected benefit of their work. Thus, through the use of survey research, the Science Citation Index, and sophisticated statistical analysis, research methodology has played an important part in the development of the sociology of science as a specialty.

Shortly after Merton's return to concentrated work in the sociology of science, there appeared two other major contributors, Thomas S. Kuhn and Derek J. de Solla Price, neither of them from sociology itself. As I mentioned earlier, Kuhn, trained as a physicist at Harvard, turned to the history of science under the influence of his teaching in the General Education science courses at Harvard sponsored by President Conant. Kuhn's first book, *The Copernican Revolution: Planetary Astronomy in the Development of Western Thought*,[27] grew out of, as he says in his preface, the "series of lectures delivered each year since 1949 in one of the science General Education courses at Harvard College." Teaching has its uses for the development of new scientific knowledge and new specialties. Kuhn sought to show in this book that science does not come just from previous science, as the now abandoned "internalist" view of the development of science held, but that the Copernican Revolution, as one extremely important development in science, was what he calls "a plurality." The core of the revolution was "a transformation of mathematical astronomy, but it embraced conceptual changes in cosmology, physics, philosophy, and religion as well."[28] Thus Kuhn showed the interrelations of the substance of science with a variety of other cultural systems. He did not yet show its interrelations with social structural elements.

This is the step that he did take in his next book, *The Structure of Scientific Revolutions*,[29] the masterpiece that has since had enormous influence on the sociology of science and far beyond. Its concept of scientific "paradigm" and its discussion of the processes of paradigm revolutions in science have resulted in a large literature. In one part of this literature, Kuhn's ideas have been applied to other cases in science than the ones he discussed and also to ideas and concepts in just about every other field of scholarship. In another part of this literature, that written by specialists in philosophy, it has also, because of Kuhn's seeming philosophical relativism, stirred a great and continuing debate. This debate concerns the essential question, Are paradigms incommensurable or not? Kuhn, it is hoped, will write a book in answer to this question.

Kuhn's book remains a fundamental focus for the sociology of scientific discovery. Indeed, Kuhn himself noticed what I called at the time in a review of his book, its "quasi-sociological" character.[30] He himself points out that "many of my [Kuhn's] generalizations are about the sociology or social psychology of scientists".[31] In that same review I remarked on the happy recent quasi-sociological development among other historians of science such as C. C. Gillispie, Hunter Dupree, and David Joravsky. This new coming together of the history, philosophy, and sociology of science in Kuhn's book, a pattern that has persisted into the present, has been valuable for all three of these specialties and not least the sociology of science.

It is worth noting for its own interest as a case for the understanding of the processes of scientific and scholarly discovery that the enormous influence of Kuhn's work was not expected either by him or by his publisher. Kuhn had promised his book to the editors of an esoteric and learned series of short publications called the International Encyclopedia of Unified Science published by the University of Chicago Press. Knowing the small sales of these paperback books, all under one hundred pages in length, the press intended to publish Kuhn in the same format. Kuhn, however, hoping for a wider audience than the series had, asked the press for a hardcover format. To support his case, he asked some of his friends in the history and sociology of science to write to the press in support of the case for hardcover publication. The press agreed, and the rest is not only intellectual history but also university-press publishing history. We see here that the sources of the development of a scientific specialty may be not only multiple and diverse but also unexpected.

Derek Price was the second major contributor, again not from sociology itself, to appear in the 1960s, almost simultaneously with Kuhn. Price had been trained in England in physics and in the history of technology. His central passion, however, was the use of mathematics and statistics for the understanding of the social aspects of science.[32]

In a review of an expanded version of Price's *Little Science, Big Science* (1986 edition), I itemized the following five contributions that I thought he had made to the development of the sociology of science:[33]

1. Price was determined to make the social and historical study of science *scientific*. He was passionate as an advocate of what he called "the science of science." Price was confident enough about his empirical generalizations to call them "laws," a scientific term with which even the most committed of sociologists and historians of science do not feel comfortable.

2. Price was outstanding for his devotion to and innovative use of quantitative data. In their brief and admiring foreword to this new volume, Robert Merton and Eugene Garfield call Price "the father of scientometrics." Price's use of quantitative data has encouraged a variety of other such use, for example, by Belver Griffith and Henry Small at Garfield's Institute of Scientific Information and by sociologists of science like the Coles.

3. Using quantitative data, Price studied the informal networks of communication and collaboration that are central in all scientific work. To characterize these networks, Price revived the seventeenth-century term "invisible college." Price's interest in network analysis coincided with an expanded interest in network analysis in sociology in general. Later work in the sociology of science by Diana Crane, Nicholas Mullins, and others has profitably used network analysis.[34]

4. Price had a brilliant writing style that dramatized the sociology of science and communicated its interest and findings widely in the scientific, scholarly, and policy-making communities. He was a master of vivid metaphors such as "invisible colleges" and the formulator of pithy aphorisms. At one point, everyone seemed to know and use his statement, "Eighty to ninety percent of all the scientists who have ever lived are alive now." Even those who did not understand what he meant by the exponential growth of science could understand that statement.

5. Price was directly concerned with the policy consequences of his findings. In the last chapter of *Little Science, Big Science* he wrote about "political strategy for big scientists." His concern was not just bookish. He was a founder and first president of the International Council of Scientific Unions.

In all these diverse and multiple ways, Price helped to shape the emerging sociology of science specialty. As a measure of his large influence, Merton and Garfield point out in the foreword to *Little Science, Big Science* that there have been 725 article references in 80 disciplines and specialties just in Price's *Little Science, Big Science*.

My own activities in the emerging sociology of science after the publication of my *Science and the Social Order* further exemplify the multiple, diverse, and unexpected origins of that specialty. My predominating sociological concern was the development of the substance and empirical application of social-systems theory, and it was in its behalf that I took up targets of opportunity in the sociology of science and also committed myself to investigating what I came to call "the dilemma of science and therapy." Several of the products of my taking up targets of opportunity are printed in part II of this book. For example, the paper (with Renée C. Fox) on serendipity in scientific discovery, "The Case of the Floppy-Eared Rabbits," was itself a result partly of chance, unexpected knowledge about the people and events described therein. Other papers were produced on special requests from organizers of symposia or from journal editors. Of course, when intensive graduate work in the sociology of science started up again at Columbia under Merton, I was able to give informal support and encouragement to the able young people emerging from his instruction.

As for my work on the dilemma of science and therapy, presented at some length in part III, that came partly out of my strong interest in the sociology of the professions and the problem of the role of values in social action, both of which Talcott Parsons had treated as among the central concerns of social-systems theory. But that work also came partly out of my wish to have an experience with the kind of theoretically relevant empirical research that I had hoped to do when I first came to Columbia University in 1952.[35]

Coinciding in the 1970s and the 1980s with important organizational developments in the maturation of the sociology of science as a specialty, there appeared an important new body of work in this field. It came primarily from Great Britain, but there were also valuable contributions from France, Holland, and Germany. It had multiple and diverse social and intellectual sources.[36]

One of the most important features of the work of the Europeans David Bloor, Barry Barnes, Harry Collins, David Edge, Karen Knorr-Cetina, Bruno Latour, Roy MacLeod, Donald MacKenzie, Michael Mulkay, Arie Rip, Steven Shapin, and Richard Whitley, among others, was their familiarity not only with the organizational aspects of science but with its substantive theories and methods. Many of them had been trained as scientists and then become researchers in the sociology of science. In some cases they knew and even still participated with the working scientists in their field, but also they were not afraid, as invited guest-researchers, to spend long periods of close observation of day-to-day, even minute-by-minute work in "strange" scientific laboratories. The result was an intimate and systematic picture of the social processes of substantive scientific discovery.

Another virtue of this European development was the way in which it brought the philosophy, history, and sociology of science into the closest and most fruitful interaction, sometimes in the work of single individuals, sometimes in the work of effective groups of colleagues, notably, for example, at the Science Studies Unit of the University of Edinburgh where its director, David Edge, has gathered and held together a set of original and productive scholars.

The origins of this development are certainly multiple and diverse. Some of the philosophers were interested in the ontological and epistemological status of scientific knowledge. Interacting with their sociology-of-science colleagues who brought to their work the concepts of "the social construction of reality" and the relativity of "discourse analysis," the philosophers and others together helped to demystify some absolutist conceptions of scientific knowledge and the processes of scientific discovery. But often they seemed to be taking an excessively relativist ontological position, asserting that the substance of science has no partial independence of its own, even if also it necessarily developed in interaction with a variety of other social-structural and cultural factors in the social system.

It was also the case that much of this work, with its heavy emphasis on the importance of what was called "interests," had something of an anti-establishment, political, ideological bias. Science, this work seemed to be saying, is not ever done in some measure for its own sake but always for some group's political, social, or economic benefit. At an extreme, some of these authors went so far as to deny the importance of values in science and in human action generally; all is group "interest" and rationalization of that interest through ideologies.

Nonetheless, after discounting the limitations I have pointed to, the numerous researches that have been done still can be counted as a valuable addition to the sociology of science. Many of their findings are valid apart from their faulty general relativist and anti-establishment assumptions. Henceforth, the sociology of science will have to include the study of both the organizational and the substantive aspects of science.[37] Truth in science does not come only out of truth; it may come out of partial error or even absolute error.

Finally, we can be brief about the final, organizational stage of the maturation of the sociology of science as a specialty, since I describe it at some length in the first paper of part I of this book. In the 1970s and 1980s the sociology of science has taken on all the characteristics of a mature scientific specialty; *it now has not only a cognitive but a professional identity.* As a specialty with a fully realized professional identity, the sociology of science now has its own journals, professional associations, regular formal national and international professional meetings, dense networks of informal interactions and footnote citations among professional specialists, honorific prizes, and specialized university courses and seminars. On the organizational side, a key occurrence in the late 1970s was the formation of the multidisciplinary professional organization, the Society for Social Studies of Science (4–S), on the initiative of some of the newer members of the specialty. They were able to gain the warm support of the senior, older members, who became its first officers, and 4–S has flourished ever since as a focus of the professional identity not only of the sociology of science but of the social science of science more generally. In sum, from multiple, diverse, and sometimes unexpected social-structural and cultural sources, the sociology of science has met with some resistance but more than enough welcome to make it a full-fledged scientific specialty. Its analytical history can teach us something of what we need to know in general about the development of scientific specialties.

## Notes

1. Barber, *Science and the Social Order* (Glencoe, Ill.: Free Press, 1952).
2. Bernal, *The Social Function of Science* (London: Routledge & Kegan Paul, 1939). Reprinted by M.I.T. Press, 1967. Also, Bernal, *The Freedom of Necessity* (London: Routledge & Kegan Paul, 1949). For a collective portrait of the scientific humanists, see Gary Werskey, *The Visible College: The Collective Biography of British Scientific Socialists of the 1930's* (New York: Holt, Rinehart & Winston, 1978).
3. Soddy et al., *The Frustration of Science* (New York: W. W. Norton, 1935).
4. Hogben, *The Retreat from Reason* (London: Watts & Co., 1936).
5. All the papers were published in a volume, *Science at the Crossroads* (London: Kniga, 1931). Hessen's paper has since been printed separately in a book using the title of his essay (New York: Howard Fertig, 1971).
6. Polanyi's most complete statement is in *Personal Knowledge: Towards a Post-Critical Philosophy* (Chicago, Ill: University of Chicago Press, 1958).
7. For an early discussion of this conflict over the planning of science between Polanyi and Bernal, including references to other relevant writings, see Barber, *Science and the Social Order*, pp. 232–37, "Can Science Be Planned?"
8. Kuhn, *The Structure of Scientific Revolutions* (Chicago, Ill.:University of Chicago Press, 1962).
9. See the four volumes of Sorokin's *Social and Cultural Dynamics* (New York: American Book Co., vols. 1–3 [1937], vol. 4 [1941]).
10. Sorokin, *Contemporary Sociological Theories* (New York: Harper & Bros., 1928).
11. Merton, "Arabian Intellectual Development," *Isis* 22 (1935): 516–24.
12. Merton, "Social Time," *American Journal of Sociology* 42 (1937): 615–29.
13. Published in *Osiris* 4 (1938), part 2, in Bruges, Belgium. Reprinted, with a new introduction by the author (New York: Howard Fertig, 1970; hardcover) and (New York: Harper Torchbooks, 1970; paperback).
14. For a partial but extensive bibliography of relevant items, see R. K. Merton, "The Fallacy of the Latest Word: The Case of 'Pietism and Science,'" *American Journal of Sociology* 89 (1984): 1091–1121. See also "Citation Measures of the Influence of Robert K. Merton," in T. F. Gieryn, ed., *Science and Social Structure: A Festschrift for Robert K. Merton* (New York: Academy of Sciences, 1980).
15. Merton, *Science, Technology and Society*, *Osiris* 4 (1938): 362.
16. Ibid., p. 414.
17. Ibid., p. 417n. Merton generalized this point in his paper, "The Unanticipated Consequences of Purposive Social Action," *American Sociological Review* 1 (1936), 894–904. This is one of Merton's most cited papers.
18. Merton, "Science and the Social Order," *Philosophy of Science* 5 (1938): 321–37.

19. Merton, "Science and Technology in a Democratic Order," *Journal of Legal and Political Sociology* 1 (1942): 115–26.
20. James B. Conant, *On Understanding Science* (New Haven, Conn.: Yale University Press, 1947).
21. Conant, *Science and Common Sense* (New Haven, Conn.: Yale University Press, 1951).
22. See also Conant's Bampton Lectures at Columbia University, published as *Modern Science and Modern Man* (New York: Columbia University Press, 1952).
23. Barber, "The Sociology of Science," pp. 226–27, in Robert K. Merton, Leonard Broom, and Leonard S. Cottrell, Jr., eds., *Sociology Today: Problems and Prospects* (New York: Basic Books, 1959).
24. Merton, "Priorities in Scientific Discovery," *American Sociological Review* 22 (1957): 635–59.
25. For Merton's studies on these and related topics, see his *The Sociology of Science* (Chicago: University of Chicago Press, 1973).
26. See, just as examples of their considerable work during the last twenty-five years, Jonathan R. and Stephen R. Cole, *Social Stratification in Science* (Chicago, Ill.: University of Chicago Press, 1973); and Harriet A. Zuckerman, *Scientific Elite: Nobel Laureates in the United States* (New York: Free Press, 1977).
27. Kuhn, *The Copernican Revolution* (Cambridge, Mass.: Harvard University Press, 1957).
28. Ibid., p. vii.
29. Kuhn, *The Structure of Scientific Revolutions* (Chicago, Ill.: University of Chicago Press, 1962).
30. In *American Sociological Review* 18 (1953): 298–99.
31. Kuhn, *The Structure of Scientific Revolutions*, p. 8.
32. See Price's two books: *Science since Babylon* (New Haven, Conn.: Yale University Press, 1961) and *Little Science, Big Science* (New York: Columbia University Press, 1963).
33. Barber, review of Price, *Little Science, Big Science . . . And Beyond*, (1986 edition), in *Isis* 78 (1987): 589–91. This book contains the original volume plus seven more papers.
34. Diana Crane, *Invisible Colleges* (Chicago, Ill.: University of Chicago Press, 1972); Nicholas Mullins, *Theories and Theory Groups in Contemporary American Sociology* (New York: Harper & Row, 1973).
35. For the fullest report on this research, see Bernard Barber, John J. Lally, Julia Loughlin Makarushka, and Daniel Sullivan, *Research on Human Subjects: Problems of Social Control in Medical Experimentation* (New York: Russell Sage Foundation, 1973). For a detailed account of the social origins and policy consequences of this work, see Bernard Barber, "The Ethics of the Use of Human Subjects in Biomedical Research (The Prototype Case)," in Bernard Barber, *Effective Social Science: Eight Cases in Economics, Political Science, and Sociology* (New York: Russell Sage Foundation, 1987). It should be clear from the fact that both of my books were supported and published by the Russell Sage Foundation that support from policy-oriented philanthropic foundations can have an important part to play in the development of a social-science specialty.
36. For a complete, admiring, but also critical account of this work, see Stephen Cole, *Social Influences on the Growth of Scientific Knowledge*, chap. 2 (Cambridge, Mass.: Harvard University Press, forthcoming).

37. Ibid. Cole, whose work has up to now concentrated on the organizational and reward-system aspects of science, here confronts the problem of the social aspects of substantive science. He takes an antirelativist stance.

**Part I**

# Historical Origins and Development of the Sociology of Science

# Introduction

As indicated in the introduction to this volume, the essays in part I overlap somewhat with the material presented therein. However, while the purpose of that introduction was primarily analytical, seeking to contribute to a general theory of the emergence, development, and maturation of scientific specialties, the essays in part I were written more for historical or honorific purposes, even though they inevitably have their analytical components. Therefore, these essays can best be taken as overlapping supplements to the analysis that is highlighted in the introduction to this book.

All of the essays were written on request for special purposes. The last of the three essays, on Sorokin's sociology of science, was written first, in the early 1960s (jointly with Robert Merton), for a *Festschrift* organized by Professor Philip Allen of the University of Virginia, an admirer of Sorokin's work as a whole. Merton and I, as students of Sorokin, were "natural" candidates to do the piece on this one aspect of Sorokin's work. The volume was published in 1963, after Sorokin's retirement from Harvard but while he was still occasionally teaching part-time at other institutions, including Virginia.

Neither Sorokin's work nor this admiring but also critical paper are much noticed now by sociologists of science. However, both Sorokin's work in the sociology of science and our paper still have their value for sociologists of science. We hope this reprinting of our paper will stimulate the attention Sorokin's work, both its virtues and its defects, deserves.

The paper "The Emergence and Maturation of the Sociology of Science" was written in 1987 as the introduction to the Chinese translation of my book *Science and the Social Order*. That translation was part of the large Chinese effort of the early 1980s to bring the Chinese abreast again of Western sociology. As I write in mid-1989, during the violent putting-down of the student democracry movement and the attendant crackdown on intellectuals, it looks as if all Chinese social science will revert, at least for the near term, to its earlier retrograde state.

Finally, my essay on Parsons and the sociology of science was the last of these three to be written (in 1989), at the request of the editors of the European sociology journal *Theory, Culture, and Society*, for a memorial issue honoring Parsons's memory on the occasion of the tenth anniversary of his death. Since his death in 1979, there has been a great increase of interest among European sociological theorists in Parsons's work. Because little attention has been paid to Parsons and the sociology of science, I chose to write my requested essay "in appreciation and remembrance" on that topic. I think Parsons's contributions to the sociology of science illuminate general features of his work and are valuable for that reason. My essay also ventures some thoughts about the relations between Parsons the man and his theories of action and of social systems. At the end, I suggest what a present-day, empirically relevant theory of social systems might look like.

# 1

# The Emergence and Maturation of the Sociology of Science

I am gratified that my colleagues Mr. Gu Xin and his associate in the Institute of Policy and Management in the Academica Sinica in Beijing, Mr. Zhao Leijin, have undertaken the translation and publication of my *Science and the Social Order* in order to make it more available to the scholarly and science-policy communities in the People's Republic of China. The book was originally published in the United States in 1952 and has since then appeared in a British edition (1953), two American paperback editions (1962 and 1970), an American hardcover printing (1978), and in Japanese translation (1955). This is, of course, the first Chinese translation.

Because, fortunately, there has been a good deal of progress in the sociology of science and in the science-policy field during the years since 1952, *Science and the Social Order* is not up-to-date with respect to data, especially those provided in the intervening years by empirical studies employing sophisticated survey research techniques and the newly invented science citation method, nor is it up-to-date in that intensive scrutiny of the substance of scientific ideas that has lately come to characterize a growing part of the sociology of science. Scholarly work carried out beginning in the 1960s and since has provided new data, new research techniques, and new intensive scrutiny of the actual development of scientific ideas. As I shall explain more fully below, I view all of this work with admiration and approval. The sociology of science has emerged and matured since 1952.

Despite all this progress, however, *Science and the Social Order* still has a great deal to offer the reader. The basic character of the book and its continuing usefulness will become quite clear in the account I give below of its intellectual origins. The purposes I had in mind in 1952 are still important for the sociology of science. The admirable progress made in the sociology of science since 1952, like all scientific progress, has often been one-sided, in this case one-sidedly microscopic in its analysis of science. It has often been neglectful of the macroscopic features of science as a major social-structural and cultural institution in society. Further, it has often not been sufficiently comparative in its approach, looking to the similarities and differences in science in different societies and in different historical periods. The social-system, institutional, and comparative presuppositions are all basic to *Science and the Social Order*, and they are still essential for a multidimensional, synthetic analysis of the social character of science.

This essay provides me a valuable opportunity to make two new statements—(1) about the intellectual origins and purpose of the book, and (2) about the welcome emergence and maturation of the sociology of science as a regular and recognized specialty in the social sciences.

### Intellectual Origins and Purpose

It is often difficult to think ourselves back to the origins of a scientific specialty. Once established, a specialty and its practitioners are more concerned with everyday, ongoing problems and work than with its origins. So it is with the sociology of science. It is difficult to realize now that in 1952 there was no such specialty as the sociology of science. True, the British "scientific humanists" (people like Bernal, Hogben, Soddy, etc.) and others in America and Russia had earlier written about the nature, problems, and uses of science, but they did not think of themselves as sociologists of science or as working in a social-science specialty. Bernal and his fellow "scientific humanists" were natural scientists concerned for the troubled world of the 1930s' worldwide depression and also for how the condition of society might be improved by the effective reorganization and greater application of science for human welfare. Their interests were primarily practical policy, not scholarship for its own sake. Those were the typical interests of the time for almost everyone writing about science and science policy.

The only professional sociologist writing about science in the 1930s and 1940s was the then-young (only in his twenties and thirties) Robert K. Merton, whose classic *Science, Technology, and Society in Seventeenth Century England*, which had been his doctoral dissertation in the new Sociology Department at Harvard, was published in 1937 under the sponsorship of George Sarton, the founding father of the history of science, in the History of Science Society publication *Osiris*. In the early 1940s, for both intellectual and social reasons (the Nazi repression of science in Germany), Merton published his also classic papers on the normative structure of science. Note that Merton did not publish either his book or his papers with a view to establishing the sociology of science as a social-science specialty. Then and now, some fifty years later, Merton's primary and overriding interest as a sociologist has been in the development of social theory. Although his book, his early papers, and his later papers, which were produced only from 1957 onward, make Merton a founding father for the sociology of science, in the 1930s and 1940s he was writing about science only because thereby he could develop theoretical ideas about the role of "ideas" or "culture" in social systems and about the consequences of ideas for stability and change in social systems. In the wholly "materialist" social theories that were frequent then, ideas were only "superstructure," only epiphenomenal, not also partly independent forces in society. Merton's thesis in his early book and papers—namely, that religious and normative ideas were independently effective in social action, both in principle and in practice (e.g., the seventeenth-century scientific revolution in England)—was intended as a contribution to social theory, a counterweight to "materialist" determinism.

As an undergraduate student of Merton's in the late 1930s, and as his associate and close friend thereafter on a continuing basis during the 1940s and 1950s, I heard nothing of the sociology of science as a specialty. Indeed, my own purpose in the late 1940s and early 1950s in setting out to write *Science and the Social Order* was quite similar to Merton's in his work on science: the development and application of social theory. We had the same purpose, but a different source for it.

As an undergraduate and a graduate student at Harvard University (with time out for four years to serve in the U.S. Navy during world War II), I had studied with and come to admire very much the contributions to social theory of Professor Talcott Parsons. I admired not only his revolutionary general theory of action but also his development of social-systems theory and his applications of that theory, in course lectures and journal articles, to the analysis of the functions, variable structures, and processes of change in social-structural and cultural systems in society. I was especially attracted by Parsons's emphasis on and exemplification of the uses of comparative historical and societal materials for the testing and development of social-systems theory. Parsons invited his Harvard colleagues who were specialists in a wide variety of societies and historical eras to give lectures in his course on comparative institutions. The basic purpose of the course as a whole and of the individual lectures by Parsons and his guests was always the development of social-systems theory.

My ambition and purpose, then, in undertaking my own first major sociological work was theoretical, to follow Parsons. I wanted to accomplish my purpose in two ways. First, I wanted to write a book that would present in concise and somewhat abbreviated form the substance of the major social-structural and cultural components of the theory of society as a social system. I intended to analyze the functions of each subsystem component within the larger social system and also its functional relations with each of the other subsystems. I intended, further, to analyze systematically the various but limited number of structural alternatives that could subserve these several social-structural and cultural functions. And finally, I intended to show dynamic processes in these subsystems, how they maintained their stability under some conditions and how they changed under others. All this was to be accomplished in a single volume with rich historical, anthropological, and contemporary empirical exemplification. I did write several chapters for a book of this kind and I did develop both an introductory and an advanced sociology lecture course in which my purpose was tentatively worked out, but no published book ever resulted. I have long regretted that fact and still think such a book would be invaluable for sociology as a science. It could provide a focus for an increase in the integration of sociology as a discipline.

The second way in which I wanted to accomplish my goal of developing and applying social-systems theory now seems incredibly grandiose. But at that time in sociology, it seemed possible. I intended to write a separate book for each and every one of the major social-structural and cultural subsystems of society. Having put aside the first—the overall, the general—book, and being familiar from my work with Parsons, Merton, and Sarton with the scattered work in the social-aspects-of-science field, I wrote *Science and the Social Order* as the *first* in the series of separate books I had in mind. In the late 1950s I wrote a second book intended both as a separate project but also as a part of my overall purpose to run through all the subsystems of the theory of society as a social system. That was my *Social Stratification: A Comparative Analysis of Structure and Process* (1957) (note the subtitle). After that, I gave up my grandiose project.

Further evidence that until the 1960s the sociology of science did not exist, nor was even conceived of as a specialty in sociology, can be found in *Science and the Social Order* itself. Although I hunted for, and found, scores of books and articles that I could organize in that book with my social-systems theory into the first comprehensive treatise on the sociology of science, when I recently examined those references I found that very few of them refer to any one of the others. We now know from recent work on the nature of specialties in science and on the invisible colleges and formal groups making them up that an essential characteristic of such specialties is the intensive cross-references of publications to one another, the co-citations, that constitute them. Specialties consist of dense webs of cross-references. By this criterion for the existence of a specialty, my book proves that no specialty existed then for the sociology of science.

A glance at the chapter titles of *Science and the Social Order* will quickly show how largely and directly indebted I am to Parsons's social-systems theory and to his work on comparative institutions. I start with a chapter on the nature, functions, and changing character of science as the essential embodiment of human rationality in social action and social systems. It should be noted that my discussion of the normative structure of science derives from Parsons's analysis of what he called the "pattern variables" as well as from Merton's classic papers on that subject. I then present comparative materials from earlier historical periods and from modern society, including the variability of science in modern society as between what I called "liberal" and "authoritarian" subtypes of modern society. Incidentally, at a time when my view was contrary to the received opinion, I argued, *directly on the basis of my social-systems analysis*, that Russian society was probably strong in many branches of science. I made this analysis at a time when one American biologist had just written a book on "the death of science in Russia." I should note that, on the basis of my analysis of Russian science, I was approached by a representative of a major U.S. government intelligence organization and asked if I would come to Washington to organize a unit for the study of Russian science. I did not go, but I am sure that the organization in question, as well as other government organizations and academic groups, now keeps up with the activities of science in all major powers, including of course, China. That science is a major source of national power is now recognized by all.

Other chapters in my book discuss organizational structures in science, first in general, and then specifically in American govenrment, industry, and university organizations. In the chapter on the social process of invention and discovery, I analyze the dynamic processes in science, concentrating on that perpetual theoretical problem in sociological theory, the relations among individual, culture, and society. The chapter on the social control of science treats of the complex, changing, and sometimes conflictual relationships between science as a social institution and other social institutions, especially, at this time and in this case, the political institution. Finally—and here I was very much influenced by Parsons's liberal optimism—I devote the last chapter to an optimistic view of the nature and prospects of the social sciences. After all these years I still hold to that view. We may not have the theoretical integration that would in many, but not all, respects be desirable for the future progress of the social sciences, but we do see a hundred flowers blooming. We are far from where I came into sociology in the 1930s. (Incidentally, when I came into sociology, I had never heard the word before. How lucky I was to fall in, as a student at Harvard College, with people like Parsons, Merton, Sorokin, and L. J. Henderson, from all of whom I learned a great deal about social theory in general and about science in particular.)

## Emergence and Maturation

This is not the place to do more than sketch some of what has happened since World War II, both intellectually and in the larger societal world, to cause the emergence and maturation of the sociology of science as a scientific specialty. In world War II, science proved, perhaps more visibly than it ever had before, its enormous social consequentiality. The development of the atomic bomb and of antibiotics, to mention only two examples, made governments see that they must give large and continuing moral and financial support to all forms of science. Postwar competition among the powers on the industrial and military fronts led to new institutions for the support of science and to the emergence of organizations and studies in science policy in the hope that they could more effectively guide such support. It is a hope that has not been fulfilled as well as some had originally wished. Nevertheless, large government support of science by all societies that can afford it is likely to be with us permanently.

On the intellectual front, a number of important developments, beginning in the late 1950s, contributed to the development of the sociology of science and science policy. At this time, Robert Merton returned from some of his intervening interests to an intensive interest in the sociology of science and encouraged a new generation of graduate students to become specialists in that field. Jonathan Cole, Stephen Cole, and Harriet Zuckerman were students of Merton's who have continued to specialize in the sociology of science. These scholars were particularly proficient in the use of survey research techniques; they thus brought new quantitative data to the sociology of science. They also made valuable use of the science citation indices that had been invented as an information retrieval device by Eugene Garfield in the 1960s but now were creatively used by sociologists to illuminate essential aspects of the social structure and processes of science.

The major part of my own work at this time concentrated on the sociology of biomedical research. In collaboration with a group of able postgraduates and graduate students (Daniel Sullivan, John Lally, and Julia Makarushka), I did two large survey studies on the problems of ethics in biomedical research. Our explicit theoretical focus was what we called "the dilemma of science and therapy." In our analysis we used our survey data, citation data, and network data. Because of the rapid heightening at his time of national-policy concern for these problems of biomedical ethics (informed consent and the risk-benefit ratio), I became involved in a series of social-policy activities and had some small influence on national social policy in this field. On the basis of that experience I have recently reflected, as a sociologist of science, on the relations between empirical social research and social policy. Using my reflections as a model for other social scientists whose empirical social research also has had policy consequences, I have collected eight cases of this kind from economics, political science, and sociology in a book, *Effective Social Science* (1987). All eight cases address themselves to the same thirty-odd sociology-of-science questions. The cases provide some interesting tentative generalizations about the relations between empirical social research and social policy.

Early in the decade of the 1960s, in *Little Science, Big Science*, Derek Price's highly mathematized publication measures of growth processes in science further promoted quantitative approaches in the sociology of science. With his revival of the seventeenth-century term "invisible college" to refer to the importance of network processes in science, Price also stimulated the use of network measures of scientific processes and structures. Other important users and developers of network measures have been Diana Crane, Nicholas Mullins, and Henry Small. Price was a passionate advocate of what he called "the science of science." His brilliant and metaphorical style made his books attractive to a wide readership.

One of the factors impeding the development of the sociology of science before the 1960s was the resistance to it by the history of science and the philosophy of science, both of which were relatively well-established scholarly specialties. In the form of an emphasis on the "internalist" approach to science, there was strong resistance in these specialties to the "externalist" assumptions of the sociology of science, particularly the Marxist version but even the more main-line, liberal version. In 1962 Thomas Kuhn's *The Structure of Scientific Revolutions* caused a creative revolution in the relations among philosophy, history, and sociology of science. Apparently not himself entirely aware of the fact at that time, Kuhn used concepts and materials from all three of these specialties in his seminal work. The study of science as a social phenomenon has never been the same since. Interspecialty cooperation and discussion is now frequent and fruitful. There is no longer any talk of "internal" and "external" aspects of science. Scientific ideas, organizations, and processes are now clearly seen by all scholars as subsystems interrelated with other social and cultural subsystems in society.

Finally, very happily, and for social and intellectual reasons that it would be most interesting to investigate as an exercise in the sociology of science, there has been another creative revolution, this one coming primarily from Great Britain. There, philosophers, sociologists, natural scientists, and science-policy specialists have combined, especially at the Science Studies Unit of the University of Edinburgh led by David Edge but also at Sussex, York, Bath, and elsewhere, to provide valuable studies of the actual substance of scientific ideas. Barry Barnes, David Bloor, Harry Collins, David Edge, Roy MacLeod, David Mackenzie, Michael Mulkay, Steven Shapin, Richard Whitley, and Steve Woolgar have been the leading British figures in this revolution.

Their close-in, micro, empirical studies of the actual development of scientific ideas or organizations are an important contribution to our understanding of science as a social phenomenon.

Like many revolutions, however, this one has its one-sided tendencies. It tends, or so it seems to me, to be relativistic on the ontological aspects of science and rationalistic on the social side. Contrary to this "school," I do not think that the natural and social worlds are entirely a construction of scientific ideas about them, nor do I think that scientific work is solely motivated by "interests," as several members of this school assert without ever giving a satisfactory theoretical definition of that term. Science is driven by science, by norms, by interests, and by the "real" world, all four.

Science has some degree of autonomy. It is to some extent *independent* of other social-structural and cultural subsystems as well as being *interdependent* with them. The difficult empirical problem is to work out specifically the amounts and types of independence and interdependence. The task is difficult, but this is no reason to shirk it.

The sociology of science has by now become an international scholarly specialty. Frenchmen such as Bruno Latour, Austrians such as Karen Knorr-Cetin, and Dutchmen like Arie Rip are all valuable workers in this specialty. The sociology of science now has its own international professional society, the Society for Social Studies of Science, and it has two professional journals, *Social Studies of Science*, founded and edited by Edge and MacLeod at Edinburgh, and *Science & Technology Studies*, the infant journal of the Society for Social Studies of Science, edited by Susan Cozzens.

The sociology of science has been born and grown up since *Science and the Social Order* was published in 1952. It is my hope that this translation into Chinese will further the progress of the sociology of science and of science policy.

# 2

# Talcott Parsons and the Sociology of Science: An Essay in Appreciation and Remembrance

Talcott Parsons made great contributions not only to the theory of action and the theory of social systems but also, and always in the light of these two presuppositional and general theories, to many subtheories and to those socio- logical "specialties," as we now call them, built on those subtheories. His contributions to such specialties as the sociology of the professions, of kin- ship and aging, religion, power and politics, and stratification, to name only some, were all seminal for those fields.[1] Of course, as is always the case in the development of science, he did not always get everything right. Who in science is ever definitive for all time? But he always asked some essential questions and gave some powerful answers that could be usefully revised, extended, improved, both empirically and theoretically as later work and time accumulated. I have myself recently tried to improve upon his theory of the professions, not by radical transformation or denial of his basic analytic start- ing points, as some work in the sociology of the professions does, but by showing his underplaying of the fact of differential power in the relations between the professional and the client, his neglect of the consequences of uncertainty and risk in the medical situation for the patient (he opened up and well understood the phenomenon of uncertainty for the doctor), and, finally, for his not making as explicit as he should have, what he understood perfectly well in theory, that the professions are motivated not only by values but also by their own interests, these values and interests usually being intertwined.[2] With these improvements, I think, his theory of the professions is still the best we have.

One specialty to which Parsons's contributions have seldom been remarked is the sociology of science.[3] In this essay of appreciation and remembrance of Talcott Parsons, I should like to call further attention to Parsons's contribution to this field. I shall do so under three heads. First, I shall speak of Parsons's passionate commitment to and interest in rationality, science, and their derivatives, experssions, and applications—all of which he felt to be absolutely essential to the character of modern society. Second, I should like to show how very much indebted my book *Science and the Social Order* (1952), which was the first attempt to state a systematic, comprehensive, comparative, and empirically based sociology of science, was to Parsons's theory of the social system.[4] And, third, I shall sketch a present-day functionalist and basically Parsonian social-system approach to the sociology of science that is, I think, an improved alternative to Parsons's own later, more abstract, hard to use empirically, version of social-systems theory.[5]

## Parsons, Rationality, and Science

Though he seemed to many who knew him an entirely cool and dispassionate man, Parsons was a man of many strong commitments, and none of these was any more passionate than his commitments to the values of rationality and individualism, both of which are basic to modern society. Because of his family background in American Protestantism and liberal social reform, because of his liberal education and activities at Amherst College, and because of his early and self-developed intellectual directions, Parsons was the quintessence of the secularized, twentieth-century version of modern man imbued with the spirit of the Protestant ethic.[6]

Parsons showed his commitment to rationality in a number of ways. For one example, in his early work on the professions, in his classic paper "The Professions and Social Structure,"[7] Parsons was concerned not only to make the general point about the importance of institutionalized norms as against individual motivation in the determination of behavior in general and in professional behavior in particular, but also to stress the essentiality in professional behavior of generalized scientific knowledge, either natural or social. Throughout his life he kept coming back to the patterns, processes, and problems of professional behavior.[8] Parsons saw the professions as the indispensable embodiments and implements of rationality and science in society. For Parsons, to take another illustrative example of his basic commitments, the university was the modern collectivity type where rationality and science were most central. In his book *The American University*,[9] co-authored with Professor Gerald M. Platt, the first chapter after the introduction is titled "The Cognitive Complex: Knowledge, Rationality, Learning, Competence, Intelligence." Although often hard to read because it was written in terms of Parsons's more abstract social-systems theory, the book (some 450 pages altogether) is a handbook of a great deal that is relevant to the sociology of science and knowledge. Parsons was always concerned not only to analyze and systematize knowledge but also to apply it. His discussion of the university, of the place in it of scientific research, of the relation of research to teaching, and of the role of teaching in the inculcation of the values of rationality and individualism, his discussion of having the professional schools in the university rather than being free-standing[10]—all this and much more are brilliantly illuminating for understanding science in the modern world and for the making of educational policy.

Parsons did not see himself as simply analyzing and understanding rationality and science in the modern world. For a man of his passionate commitment to these values and activities, it was equally important to promote them. This he did in a number of ways. For one example, in the late 1940s, when the American government was considering the establishment of what eventually became the National Science Foundation and there was much opposition from some powerful natural scientists to the inclusion of the social sciences in such an organization, Parsons wrote, at the request of the Social Science Research Council, an eighty-page monograph, *Social Science: A Basic National Resource*, presenting the case for the social sciences. For political reasons, this essay was not published at the time but has finally come into print.[11] At the time it was being written, I was a tutor and teaching fellow working with Parsons in the newly founded Department of Social Relations at Harvard. Parsons asked me to read and comment on it; thus I became thoroughly familiar with it. As will be reported more fully below, Parsons's monograph was most useful to me in the subsequent writing of *Science and the Social Order*, not least of all in my optimistic last chapter, "The Nature and Prospects of the Social Sciences." As a committed promoter of science in general and social science in particular, Parsons was, to the end of his days, an optimist about the development of social science.

No promotional task was too small for Parsons. After he had been honored by election in 1949 to the presidency of the American Sociological Society and had given his optimistic presidential address, "The Nature and Prospects of the Social Sciences,"[12] Parsons volunteered to take on the society's routine, workaday administrative position as secretary. Nearer to home, he was not only the intellectual founder of the Social Relations Department at Harvard, planning therewith to realize his hopes for a truly integrated, interdisciplinary social science, but he served conscientiously and effectively as its chairperson for several years. And here is one last example of his willingness to promote the social sciences. In the mid-1960s he accepted an invitation from the United States government's agency, the Voice of America, to organize a series of radio talks on American sociology for its Forum series, to be broadcast throughout the world.[13] Parsons succeeded in inducing twenty-three distinguished American sociologists representing the various schools of American sociology, some Parsonian, some not, to prepare talks on as many topics. In his introduction to the published volume of these talks, Parsons made it clear that the essays were intended for "nonprofessionals" and that they added up to what he thought represented what the economists of development had come to call a "take-off" point in the development of sociology. Here, as always, it was Parsons the optimist and promoter speaking.

The preceding discussion of rationality and science as both values and activities may raise for some readers of this essay the question of what Parsons believed about the relationship between values and science, that is, science in its systematic substance as a set of theoretical ideas. Did he hold a so-called value-free position on the status of the substance of science? Did he believe that it was not at all influenced by values? Or did he think, on the contrary, that values determined the substance, that science, to phrase what might be thought to be an "idealistic," upside-down version of Marxism, was epiphenomenal to values? Parsons held neither of these mistaken positions. For him, both scientific theories and values were partly independent, partly interdependent sybsystems in the larger social system of society. That is to say, each influenced the other in specifiable ways, but each also had its degree of functional independence. Parsons was no ontological relativist, no absolutist on the issue of "the social construction of reality." In the final section of this essay, when presenting what I think is an improved present-day version of social-systems theory for the sociology of science, I shall discuss in a generalized way this issue of the independence and interdependence of subsystem variables in a generalized theory of social systems.

## *Science and the Social Order* as an Exemplar of Parsonian Social-Systems Theory

At the beginning of this essay, I stressed the importance of Parsons's contributions to social-systems theory in addition to his contributions to action theory. These latter contributions have been much more discussed in recent comment and criticism on Parsons's work. This is, it seems to me, one sign of the desirable return of sociological theory to presuppositional basics.[14] In this section of my essay I shall be dealing primarily with Parsons's social-systems theory.

I can report from personal experience and from course notes in my possession, that already in the middle 1930s Parsons was presenting at least the solid beginnings of a systematic, comparative, empirically based social-systems theory. In 1935 I entered Harvard College as a freshman, never having heard the word "sociology"; that was common in those days. Because of previous reading, interests, and part-time work on a local newspaper in Cambridge, Massachusetts, where I lived, I intended to use my college education to become either a lawyer or a journalist (I was attracted to the latter by my reading of the star journalist Vincent Sheean's recently published and best-selling book, *Personal History*). In my freshman year, therefore, I took introductory courses in history, government, economics, and philosophy, but not in sociology. Fortunately, as it turned out, a friend recommended that in the following year I take an introductory course in sociology. It was a marvelous introduction from Pitirim Sorokin, who gave two lectures a week, and from Robert Merton, who gave a third lecture-discussion session as Sorokin's course assistant. As a result of that course and also by discovering that in my tutorial reading that year as a major in government, I much preferred the more sociological works of people like Thorstein Veblen and Harold Lasswell to those of straight political scientists, I went for sociology. (There was little political sociology then.) I gave up my government major for a major in sociology and a tuteeship with Robert Merton. That was fifty years ago and I have never looked back from sociology.

In my junior year, I took Assistant Professor Talcott Parsons's course, Social Institutions, in which he had been developing for some years his social-systems theory: society was a system composed of such interrelated social-structural and cultural subsystems as stratification; kinship; the economy and property; the state, politics, and authority; language; art; science; and religion and magic. All of these weekly and multiple-weekly topics were illustrated by lectures and reading from such diverse societies as ancient China and India, Western feudalism, the Navajo, Victorian England, the Greek *polis*, Antonine Rome, and the Ottoman empire. (Parsons had not yet closely studied Freud so there was only passing mention of personality factors interacting with social-structural and cultural ones; not yet the intensive theoretical and empirical discussion that was to come in his later work.)

How could one person have expert knowledge of so many different bodies of historical and comparative materials? Parsons of course didn't have such knowledge, but, showing the interdisciplinary outreach that he always had, he invited expert colleagues from all the different relevant departments of the Harvard faculty to lecture and recommend readings on their subject specialties. Through prior discussion with his colleagues, and through their continued cooperation over the years, Parsons more or less successfully conveyed to them what was valuable for his social-systems theory. It was a feast of a course, both theoretically and substantively.

Of course, Parsons didn't slight explicit theoretical analysis. Indeed, he devoted his first lecture to pointing out the indispensability of theory, or conceptual schemes, for all science.

Parsons gave a shortened version of his Social Institutions course in the late 1940s in the newly established Social Relations Department, but it was not long before he moved away from this more empirically based analysis to his more and more abstract AGIL, four-function theoretical system with its large-scale multiplication of major, minor, and subminor theoretical categories. It was this later social system that Parsons' critics derided as "the manufacture of empty boxes," asserting that Parsons, the self-styled "incurable theorist," care more for "mere" theory than for its empirical use and relevance. However, as I have recently argued, Parsons was extremely sophisticated in the logic of theory. He very well knew in principle that theory must be empirically relevant, but in practice he came more and more to stress the development of abstract theory as against empirical illustration and research data.[15] He never ceased to feel that his most abstract theorizing was valuable for empirical analysis. He knew what the goal was even when he was not reaching it too successfully.

It is Parsons's earlier version of social-systems theory that has been the main focus of my own work for some forty years. In my teaching and writing, I have tried to extend it both through actual use and through direct theoretical consideration. Of course, I learned also from Parsons's later work, both his abstract theorizing and his more empirically focused essays, but I came to find my own development and use of his theory more satisfactory.

How does all this relate specifically to my *Science and the Social Order?* In 1948, after four years of naval service in World War II and two more years at Harvard, I left for my first independent teaching position. I intended my major scholarly contribution to be the development and application of the social-systems theory I had learned from Parsons. With a sociologist friend, I set up what now seems a most grandiose project, although it is perhaps a sign of the development of sociological theory, research, and methodology since then that at the time it did not seem so grandiose. Basically following Parsons's earlier social-systems theory and its development, my friend and I were to write a volume laying out concisely, briefly, with full empirical illustration from comparative materials, the model of society as a social system. The volume was to include a discussion of both stability and change in social systems. This "outline" volume was to be followed by a series of volumes on each of the social-structural and cultural subsystems of society. As I said, a grand project.

Unfortunately, for personal reasons, not substantive ones having to do with the design of the project, our plan was not realized. Regrettably, the "outline" volume was abandoned after some seven chapters had been written. I still like to think that a volume of this kind could have had a most beneficent effect on the development of recent sociology.

I then turned to the second part of the plan, to write, in social-systems terms, a volume on each of the social-structural and cultural subsystems in society. I chose to start with science. Because the book I wrote *Science and the Social Order* (1952), was the first systematic, comprehensive, and empirically based book on the sociology of science; because it has appeared in several editions and languages since then; and because it has recently been declared a "citation classic" by the Science Citation Index,[16] it has been widely and continuously assumed that the book was written as a contribution to the now well-established sociological specialty, the sociology of science. It is clear from what I have said above that this assumption is wrong. Only in the 1970s and since has the sociology of science become a visible and established specialty, with its own professional associations (national and international), journals, conventions, prizes, and all the other apparatus of a specialty. For example, the Sociology of Science section in the American Sociological Association was established only in 1988. Before there was a recognized specialty, there was brilliant and seminal work by Robert Merton;[17] and in the 1960s and thereafter he trained important students in the field, but certainly I had no idea in 1952 of contributing to a specialty. I was contributing an exemplar to social-systems theory. I should have said so explicitly.

Why did I choose science as against all the other equally good possibilities? Very simply, it was because of my past training. As an undergraduate and graduate student at Harvard, I had come to know, through quite diverse course work, reading, and discussion, a variety of people who worked in the embryonic fields of the history and sociology of science. In the latter, of course, there was my friend Robert Merton. In the former, there was President James Bryant Conant of Harvard, Professor L. J. Henderson, George Sarton, I. B. Cohen, and Giorgio de Santillana. With this background, it was easiest for me to turn first to a book on science.

The imprint of Parsons and his early social-systems theory is all over my book. The first chapter, as might be expected, is titled: "The Nature of Science: the Place of Rationality in Human Society." As Parsons would have, and with direct debts to his lectures and to his Social Science Research Council (SSRC) monograph on behalf of the social sciences, I argued the independent character of human rationality and science about the natural and social worlds. In describing the character of science, I stressed the key importance of its systems of theory, using the term for such systems that we had learned from Conant, "conceptual schemes."[18]

In my second chapter, I moved on to illustrate the interdependence of science with other subsystems of the larger social system, emphasizing their reciprocal effects on one another, criticizing one-factor theories such as Marxian theory. Some of the other factors I mentioned were political forces, philosophical assumptions, religion, and economics. I took my illustrations from Greek society, feudalism, the early modern and modern periods of Western society, and primitive societies. Definitely Parsonian all the way.

In chapter 3 I undertook to discuss, more intensively still, "Science in Modern Society: Its Place in Liberal and Authoritarian Societies." Here I urged "the special congruence" of science with several characteristic subsystems of modern "liberal" societies such as the United States and Britain and its difficulties in the face of certain characteristics of such "authoritarian" societies as Nazi Germany and Stalinist Russia. Incidentally, I was careful to point out that in Soviet Russia, unlike in Nazi Germany, there was a mix of favorable and unfavorable social conditions for science. This meant that I was less surprised than many other Americans when the Russians sent Sputnik aloft in 1957. (As a result of my relative objectivity about Russian science, I was visited by an official of an American intelligence agency—probably the C.I.A. but not identified by him—and invited to organize studies of Russian science for the American government in Washington. I did not take up this offer.) In this chapter I also devoted much attention to the importance for science of a certain set of values, building my analysis primarily on Robert Merton's classic paper but also on Parsons's first statement of the pattern-variables, in his professions paper.[19] Little wonder that at the beginning of this chapter I quoted the following from Parsons's then unpublished "Social Science: A Basic National Resource":

> Science is intimately integrated with the whole social structure and cultural tradition. They mutually support one another—only in certain types of society can science flourish, and conversely, without a continuous and healthy development and application of science such a society cannot function properly.

Other chapters where Parsons's influence was direct and beneficial were chapter 6, "The Scientist in the American University and College"; chapter 10, "The Social Control of Science"; and Chapter 11, "The Nature and Prospects of the Social Sciences," which, as I have indicated earlier, took the same optimistic stand that Parsons did on this matter.

In sum, *Science and the Social Order* is very much an exemplar of the application of Parsons's social-systems theory. Because there has been a good deal of progress since 1952 in the sociology of science and science policy, the book is not up-to-date with respect to data, especially from those empirical studies provided in the intervening years by sophisticated theory-cum-survey research and the newly invented science citation method, nor is it up to date in that intensive scrutiny of the substance of scientific ideas that has recently become widespread.[20] I view all of this work with approval and admiration. The sociology of science has emerged and matured as a highly visible and established sociological specialty since 1952. Nevertheless, despite its shortcomings, *Science and the Social Order* is still worth study as an exemplar of Parsons's social-systems theory and the empirical use thereof. Its emphasis on the larger societal-system and comparative contexts of the more microscopic structures and processes of science that have been a heavy focus of present work is still essential for a satisfactory sociology of science.

## A Sketch of a Social-Systems Theoretical Approach to the Sociology of Science

I said in the introduction to this essay that I wanted, in this third section, to sketch a present-day functionalist and basically Parsonian social-systems approach to the sociology of science that is an improved alternative to Parsons's own later, more abstract, hard-to-use-empirically version of social-systems theory. The project for developing a usable *generalized (abstract), systematic, and comprehensive* theory of social systems is still *one* of the most important tasks for sociological work. For forty years, through the development of the necessary theory and through continuous empirical use, which has in turn been essential for the development of the theory, starting from Parsons's early work on social-systems theory, I have myself been working on this project. In this third section, for lack of space, I present only the briefest methodological and substantive statement and an accompanying model of the theory.[21]

There are at least three characteristics of present-day sociological theory that require improvement in the light of such a theory and model. The first is the *vagueness* and *diversity* of many of the terms in current use. Terms such as "science" itself, "ideology," and "culture" are used in multiple, vague, and conflicting ways.[22] An essential function of a usable social-systems model would be to specify and standardize, not for all time but at least provisionally, a set of theoretical, empirically relevant categories.

The second unfortunate characteristic of much theory and research is its *reductionist and absolutist* quality. All causative analysis is reduced to a single variable and that variable is then absolutized as the one and only determinant of all behavior. There has probably not been a single social-structural, cultural, or personality variable that has not in someone's work been offered in this reduced and absolutized form. The virtue of a social-systems theory in which not only the broad categories of social structure, culture, and personality but also the subsystem categories within each of these broader categories are assumed, as a basic presuppositional matter, to be partly independent of one another, partly interdependent with one another, is that we can use it to explore the various kinds of independence and also the multivariate interdependencies that are necessary for an accurate analysis of concrete behavior.

Third, the lack of a *generalized, systematic, and comprehensive* theory and model for the social system leads to the ad-hocism that is rampant in present-day sociological work. In a great deal of our work, there seems to be no theoretical rhyme or reason for the selection of the variable or variables used; they seem to have been selected ad hoc, because of available data or past training or ideological disposition, or whatever. They are not located in a theoretical context such as is provided by an explicit generalized, systematic, and comprehensive theoretical model. Nor can they contribute to the development of such a model, as they ought to.

These three faults in present-day sociological work would be eliminated by the use of a model that presupposes the following: the independence and interdependence of all its categories; a functional or causative relationship among its categories; and the possibility of both stability and change in the available alternatives within and among the categories. It is an *inherently provisional* model, as all scientific theories and models should be. For example, in recent years it has become apparent that the social-structural category of gender should be added as a separate and partly independent (but not absolutized) variable. Finally, the words "Avoid 'the fallacy of the list'" should be added to the model to signify the theoretical principle of the equality of all the category variables; no one is theoretically to be preferred over all the others; the variables can be listed in any order so long as they are all listed.

So much for an excessively brief general statement. How does the model apply specifically to the sociology of science? Here are just a few points, by way of illustration. In one respect, science is a part of "culture," is a set of ideas, of symbol systems, with some degree of independence, the absolutist and ontologically relativist social-constrution-of-reality school to the contrary notwithstanding. But, in other respects, concrete science is analytically a stratification and reward system (as the Mertonian model and work has demonstrated); it is influenced by cultural values and religious ideas (as again Merton's work has demonstrated); and it is carried on, communicated, and taught in characteristic forms of organization. Indeed, when we look at concrete science clearly, we see that we need all the social-structural, cultural, and personality variables stated in the model to provide a satisfactory analysis of its nature and development. That is what I was aiming at, with only partial success, in *Science and the Social Order*. With the present model in mind, I could do better today than I did in 1952. And, of course, I think the model would be equally superior if sociologists started with other variables than science. But that is another story than the one I have focused on here. Some version of Parsons's project for a usable generalized, systematic, and comprehensive theory of social systems must come to prevail in sociological theory and research.

## Notes

1. For exemplars, see any of the several collections of Parsons's essays.
2. Bernard Barber, "Beyond Parsons's Theory of the Professions," in Jeffrey Alexander, ed., *Neofunctionalism* (Los Angeles: Sage Publications, 1985), pp. 211–224.
3. But, for some recent notice, see Barber, "The Emergence and Maturation of the Sociology of Science," in *Science and Technology Studies* 5, nos. 3/4 (1987);

129–33; in connection with this article, see also Barber, "Correction," *Science, Technology & Human Values* 13 (1988): 215. And also see Samuel Z. Klausner and Victor M. Lidz, eds., *The Nationalization of the Social Sciences* (Philadelphia: University of Pennsylvania Press, 1986). Parsons's contribution to this volume consists of an essay written in 1948 and previously unpublished. It is further described later in this essay.

4. For the citations to this edition and several later ones, including a Chinese translation now in press, see Barber, "Correction."

5. This alternative is much more fully presented in a book in progress, *The Meaning of Culture*.

6. Parsons, as a very young man, translated Max Weber's *The Protestant Ethic and the Spirit of Capitalism* (London: Allen & Unwin, 1930). For an account of Parsons's formative family and educational background, which is based on work in the Parsons's papers at Harvard and on interviews with family members, see the work in progress by Dr. Bruce Wearne of the Chisholm Institute, Melbourne, Australia.

7. See *Social Forces* 17 (1939): 257–67.

8. Again, see several of Parsons's collections of papers.

9. Parsons and Platt, *The American University* (Cambridge, Mass.: Harvard University Press, 1973). In the "epilogue" to this book, Professor Neil Smelser, of the University of California, Berkeley, a one-time collaborator of Parsons and lifelong intellectual colleague and friend, presents some valuable Parsonian amendments to the main Parsons-Platt analysis. This is the kind of useful revisionism that Parsons's work lent itself to.

10. See especially chap. 5, "The University and the Applied Professions: The Professional Schools."

11. It forms chap. 2 in Klausner and Lidz, eds., *Nationalization of the Social Sciences*.

12. Parsons, "The Nature and Prospects of the Social Sciences," *American Sociological Review* 15 (1950): 3–16.

13. Eventually published as *American Sociology: Perspectives, Problems, Methods* (New York: Basic Books, 1968).

14. See, e.g., Jeffrey Alexander, *The Modern Reconstruction of Classical Thought: Talcott Parsons*, vol. 4 of Alexander, *Theoretical Logic in Sociology* (Berkeley: University of California Press, 1983).

15. See Barber, "Theory and Fact in the Work of Talcott Parsons," in Klausner and Lidz, eds., *Nationalization of The Social Sciences*.

16. See *Current Contents: Social and Behavioral Sciences* 20, no. 27 (1988).

17. See R. K. Merton, *The Sociology of Science: Theoretical and Empirical Investigations* (Chicago, Ill.: University of Chicago Press, 1973). This is a collection of Merton's work from the 1930s onward.

18. For Conant's important work, see, for example, his *On Understanding Science: An Historical Approach* (New Haven, Conn.: Yale University Press, 1947). Thomas Kuhn has acknowledged the importance for his own ideas of Conant's work and colleagueship.

19. See Merton, "Science and the Social Order," *Philosophy of Science* 5 (1938): 321–37. Parsons, "The Professions and Social Structure," *Social Forces* 17 (1939): 257–67.

20. For ample illustration of the progress made in all three respects, see the almost twenty published volumes of *Social Studies of Science: An International Review*

*of Research in the Social Dimensions of Science and Technology*, edited by David Edge and Roy MacLeod at the Science Studies Unit in Edinburgh University. See also, *Science, Technology and Human Values*, now the official journal of the Society for Social Studies of Science, the international sociology of science association.

21. For a complete statement, see Barber, *The Meaning of Culture*, cited above as a work in progress.

22. For an intensive discussion of this vagueness and diversity with respect to "ideology," see Barber, "Function, Variability, and Change in Ideological Systems," in Bernard Barber and Alex Inkeles, eds., *Stability and Social Change* (Boston: Little, Brown, 1971).

# 3

# Sorokin's Formulations in the Sociology of Science

*Robert K. Merton and Bernard Barber*

From the beginning we must abandon the attempt to put into short compass all the wide-ranging, diversified, and developing observations in the sociology of science set forth by Pitirim Sorokin. Any such effort would be the work of a sizable book, not of a short essay. For his contributions to the sociology of science engage almost every other major part of his empirically connected sociological theory. To try to trace out each component in his sociology of science  or, more generally, in his sociology of knowledge—would mean to touch upon every other aspect of his voluminous works. In place of a systematic treatment of Sorokin's contributions to the sociology of science, therefore, we shall substitute some observations that bear upon their most significant and sometimes thorny aspects. In place of the many details that enter into his sociology of science, we shall put the more general formulations that encompass these details, knowing that this means the exclusion of issues that in a more thorough examination would have to be taken up substantially. And finally, in place of tracking down the development of Sorokin's ideas about the sociology of science as these emerged over the course of almost half a century, we shall deal primarily with his later ideas, particularly as these were set forth in his *Social and Cultural Dynamics*. In short, this is only an essay toward a critical understanding of Sorokin's work in the sociology of science; it is not a comprehensive and methodical analysis of that work.

As though this were not enough of a limitation, we must confess to another. We find it difficult, not to say impossible, to achieve a sufficient sense of historical distance from a scholar of our time and a scholar, moreover, who has been the teacher of us both. But here Sorokin himself has come to our rescue. And the respect in which he has done so is itself almost an essay in the microsociology of science, showing how the social structure of a university affects the relations between a professor and his students and so affects the transmission of knowledge. Sorokin first began to make his imprint upon American sociology shortly after he took up his first post in the Untied States at the University of Minnesota. But it was especially after he was brought to Harvard to found the Department of Sociology there that this emphatic, straight-spoken, and *urgent* man began to influence the thinking of substantial numbers of students. In order to understand the character of this influence, we should try the thought-experiment of imagining that Sorokin had gone not to Harvard but to some other important university—in Europe. He would have held "the chair of sociology" in that university. He would have had a number of assistants as well as students who, in accord with the cultural expectations of docility, would have become his disciples, echoing his words and thoughts almost as though they were their own. But in the American scheme of things academic, and particularly in a university such as Harvard, the social structure and the culturally defined patterns of expectations were of course quite otherwise. The authority-structure of the department was pluralistic rather than strongly centralized. There was not only *the* occupant of The Chair but other members of the faculty who had equal access to students. Nor was there an unchallenged norm of obedient agreement with the major professor. This structural situation meant that, at Harvard, many graduate students in sociology were apt to become anything but disciples. Moreover, Sorokin's own personality and role-behavior reinforced this tendency toward independence of mind among his students. They tended to adopt the same critical stance toward aspects of Sorokin's work as he, in the capacity of a role-model, was taking toward the work of others, both contemporary and by-gone. All this meant that the structurally defined role of Sorokin was primarily that of alerting students to intellectual alternatives rather than that of imprinting the particulars of his own theory upon them. And all this, we conjecture, helps explain how it is that Sorokin's students have not hesitated to differ with him when, rightly or wrongly, they did not see matters just as he did. In adopting this sometimes timorous but socially supported position of criticism, they were helped in no small measure by the prototype of Sorokin's own behavior, when he was persuaded that other scholars had erred. This, then, is no occasion for exhibiting alumnal piety in public but rather an occasion for applying to Sorokin's work the same critical standards that he has applied to the work of others. It is primarily an essay in criticism rather than an essay in exposition.

### 1. Sorokin's Sociology of Science: The Central Position

Sorokin has explicitly adopted an idealistic and emanationist theory of the sociology of science. Unlike the theories of a Marx or a Mannheim, which seek primarily to account for the character and limits of knowledge obtaining in a particular society in terms of its social structure, Sorokin's theory tries to derive every aspect of knowledge from underlying "culture mentalities."[1] So prominent is this aspect of Sorokin's theory that it has been variously noted by every commentator upon his work. In somewhat restricted terms, for example, it has been observed by Maquet, a thoroughly sympathetic critic, that in Sorokin's theory the "independent variable is the intellectual position in regard to ultimate reality and ultimate value. . . . The three premises of culture are nothing else but philosophic positions."[2] This statement, as we shall soon see, places far too restrictive an interpretation upon the theory. By Sorokin's own testimony, much more is contained in his concept of types of culture than can be aptly described as the philosophic position basic to it. Nevertheless, Maquet does take hold of the essential fact that each of the types of culture discriminated by Sorokin has its distinctive ontological orientation. This has also been noted, to cite only one other commentator on Sorokin, by Stark, who observes that "It is essential for the understanding of the whole theory to realize that it considers the ontological convictions prevailing at a given time not so much as culture contents but rather as culture-premises, from which the culture proceeds and emanates as a whole."[3]

What, then, is the character of the "culture mentalities" that are variously expressed in distinctive kinds of (claims to) knowledge and how do these mentalities differ? To answer this question in detail is not possible here. But enough must be said here to reidentify the core of Sorokin's conceptions. There are then two "pure types" of culture mentalities, which differ fundamentally in what is taken to be the nature of ultimate reality and value. The first is the ideational, which conceives of reality as "non-material, ever-lasting Being," which defines human needs as primarily spiritual and seeks the satisfaction of these needs through "self-imposed minimization or elimination of most physical needs."[4] The ideational culture adopts the "truth of faith." At the other extreme is the sensate mentality, which limits reality to what can be perceived through the senses. Concerned primarily with physical needs, this culture calls for satisfaction of these needs, not through modification of self, but through modification of the external world. It is oriented to the "truth of senses." Intermediate to these two is a "mixed type" of culture mentality, the idealistic, which represents a kind of balance of the foregoing types. It is oriented toward a "truth of reason." From these three types of mentalities—the major premises of each kind of culture—Sorokin derives their distinctive systems of truth and knowledge.

We shall have something more to say about the meanings, both expressed and implicit, of the idea that these culture premises are basic to distinctive kinds of knowledge developing within each kind of sociocultural system. For the moment, however, we need note only that Sorokin takes a great array of somewhat more specific types of knowledge as *dependent* upon these cultural premises. These include such dependent "variables" as the fundamental categories of causality, time, space, and number; basic philosophical conceptions such as idealism-materialism, eternalism-temporalism, realism-conceptualism-nominalism; various conceptions of cosmic, biological, and sociocultural processes, as expressed, for example, in notions of mechanism or vitalism in biology; the rate of scientific advance; the prevailing kinds of moral philosophy; and to take just one other, the various kinds of criminal law.

Sorokin himself best summarizes the brooding omnipresence of the three principal types of culture:

> Each has its own mentality; its own system of truth and knowledge; its own philosophy and *Weltanschauung*; its own type of religion and standards of "holiness"; its own system of right and wrong; its own forms of art and literature; its own mores, laws, code of conduct; its own predominant forms of *social relationships*; its own economic and political organization; and finally, its own type of *human personality*, with a peculiar mentality and conduct.[5]

In short, more than the forms of "knowledge" alone are dependent upon the premises underlying each type of sociocultural system. The forms of social structure and the kinds of prevailing personality also share this condition of dependence. Every sphere of culture, social structure, and personality is seen as emanating from the fundamental orientations characteristic of each of the three kinds of sociocultural systems.

As we shall also see in due course, Sorokin considers that particular theories of science as well as the rate of scientific advance are dependent upon these underlying cultural premises. Here we need note only the first of puzzles presented in Sorokin's conception, which, so far as we can see, he does not solve for us. How can he escape from the self-contained emanationism of the theoretical position he adopts? For it would appear tautological to say, as Sorokin does, that "in a Sensate society and culture the Sensate system of truth based upon the testimony of the organs of senses has to be dominant."[6] For, sensate *mentality*—that abstraction which Sorokin makes ontologically basic to the culture—has already been *defined* as one conceiving of "reality as only that which is presented to the sense organs."[7] In this case, as in other comparable cases, Sorokin seems to vacillate between treating his types of culture mentality as a defined concept or as an empirically testable hypothesis. This is the first of several questions which it would be useful to have Sorokin examine anew and clarify for future reference.

By way of introduction, then, we see that Sorokin's sociology of science, and by extension, his sociology of knowledge, is idealistic and emanationist. In this respect it differs fundamentally from the materialistic conception of Marx, which focuses on the intellectual perspectives generated by the position of thinkers in the class structure of their time, just as it differs from the quasi-Marxist conceptions of Mannheim, which extend Marx's notion of the structural bases of thought. It is this contrast, presumably, that led Maquet to conclude that "since Sorokin's independent variable is the intellectual position in regard to ultimate reality and ultimate value, his sociology will have a very idealistic character (in the current sense of the word . . . 'ideas rule the world'). . . ."[8]

This can be put somewhat differently without altering the basic point, but directing attention to some of its further implications. In one of his many books published after the *Social and Cultural Dynamics*, Sorokin addressed himself to the tripartite distinction of culture, social structure, and personality as abstracted aspects of all human action. This can be seen from the title of the book: *Society, Culture and Personality*. Although he attends to all three aspects, even in his earlier work, it is plain that in his sociology of science Sorokin asserts the dominance of culture over the other two aspects of social structure and personality. It is being asserted, apparently, that deep-rooted cultural assumptions override any variation in social structure and in diverse types of personality, producing a basic uniformity of outlook that is characteristic of people living in a particular kind of culture. Cultural mentality is regarded as fundamental; social structure and personality as producing, at most, minor variations on culturally embedded themes.

There is much to be said for this position, *providing* that it is not allowed to become a barely disguised tautology. To the extent that people in a particular society do in fact share the same fundamental assumptions about a reality significant to them, to the extent that they share much the same values, they will indeed tend to express this in their behavior and in their works. But the extent to which they do share these orientations and values is of course a question of fact, rather than an assumption of concept. It is a matter for inquiry, not a matter of conviction, to find out how far this obtains in different cultures. In short, a major problem of inquiry is to find out how far there obtains a consensus of outlook and to explain the differences that are found to exist among people living in the "same" culture. Sorokin has addressed himself to this question, but only in part. His data, as we shall now see, demonstrate substantial variability *within* each particular type of culture, but Sorokin's theoretical commitments and preferences are such that he does not go on to examine some of the implications of this variability in order to extend and refine his theory of the sociology of science. In a word, it is being said that Sorokin has confined himself to a first approximation—to be sure, a vast and comprehensive one—but one which sets excessive limits to a sociology of science that must attend also to the bases of the considerable variability of scientific outlook within the same culture. The different conceptions set forth within the same culture, it is being suggested, need not be merely minor variations on a major theme; some of these variations are precisely those that lead to basic developments in thought and science. Committed to the first approximation—to the focus on dominant tendencies within a culture—Sorokin largely shuts himself out from *analyzing* those variations, which often make for the advancement of cumulative knowledge, that which is recognized as authentic knowledge in cultures that otherwise differ in many respects.

This statement is a large claim. At the least, it must be elaborated if we are to assess the limitations as well as the contributions of Sorokin's sociology of knowledge in general, and of his sociology of science in particular.

## 2. Macro- and Microsociological Perspectives on Knowledge

Sorokin's theory of sociocultural systems is offered as a partial description and analysis of the whole vast sweep of Greco-Roman and Western societies—with something more than a casual orientation to Eastern societies—during the last three thousand years or so. The theory may fairly be described as a macrosociological perspective.[9] It attends to the gross rather than the microscopic features of each society and culture under view. The centuries-long periods of culture described as ideational, idealistic, or sensate, for example, are characterized in terms of their dominant traits. Discriminations within each culture are largely excluded by the very scope of the conception. This exclusion, we must realize, is imposed by the theoretical commitment, not by the external reality. Concretely, it means, for example, that for Sorokin the last four or five hundred years in the West comprise a singly sociocultural type dominated by a sensate culture mentality. All the substantial variations of science and knowledge generally that are to be found in this period are, under Sorokin's comprehensive scheme of analysis, regarded as expressions of one fundamental orientation toward reality.

Now, first approximations in approaches to social reality have a way of concealing, deep within them, basic commitments to values. For what is singled out as *fundamental* is what the observer takes to be that which "really matters." And in the same obvious sense, the variations that are excluded from notice by the observer's conception are thereby regarded as inconsequential for what the observer regards as significant. So it is that by characterizing the entire period as sensate, Sorokin does not direct his analytical attention to the differing kinds of scientific work that are to be found in that uniformly sensate period. This is a perspective which, precisely because it is macroscopic, throws together, for all pertinent purposes, the work of a Galileo, Kepler, and Newton, on the one hand, and the work of a Rutherford, Einstein, and, shall we say, Yang and Lee, on the other. It thus excludes from analysis the great differences that, for many human and intellectual purposes, are to be found in the science of the sixteenth or seventeenth century and the twentieth. It is a gross approximation that threatens to usurp the attention of those who have reason to regard the variability *within* the macroscopic sensate period as also fundamental.

These remarks may be enough to raise the second question that, in our view, needs to be dealt with by Sorokin in reexamining his sociology of knowledge and of science. What components of his theory will help us to account for the variability of thought within the societies and eras which Sorokin assigns to one or another of his sociocultural types? Has Sorokin imposed excessive limits upon his theory by a commitment to a kind of macroscopic analysis that excludes from detailed investigation the very questions that many would consider central?

In one restricted sense, Sorokin does address himself to this problem. He observes, for example, that the failure of the sensate "system of truth" (empiricism) to monopolize our sensate culture testifies to the fact that the culture is not "fully integrated." But this would seem to surrender inquiry into the bases of those very differences of thought with which our contemporary world is concerned. On the Sorokinian theory, how does one account for these differences? The same question applies to other categories and principles of knowledge with which he deals on the plane of macrosociology. He finds, for example, that in present-day sensate culture, "materialism" is less prevalent than "idealism," and that "temporalism" and "eternalism" are almost equally current, as are "realism" and nominalism," "singularism" and "universalism." And now we come again to the decisive (and we repeat, self-imposed) limitation of Sorokin's theory: since, by his own testimony, these diverse doctrines exist within the same culture, how can the general characterization of the culture as "sensate" help us to explain why some thinkers subscribe to one mode of thought, and others to another?

The essential point is that Sorokin's theory does not lead us to explore variations of thought *within* a society or culture, for he looks to the "dominant" themes of the culture and imputes these to the culture as a whole.[10] Quite apart from the *differences* of intellectual outlook of different classes and groups, contemporary society, for example, is regarded by Sorokin as an integral example of sensate culture. On its own major premises, Sorokin's theory is primarily suited to characterize cultures in the large, not to analyze the connections between various positions in the social structure and the styles and content of thought which are distinctive of them.[11]

That the macrosociological level of analysis excludes from attention problems that are of import in understanding varied developments within a culture has been noticed by several critics of Sorokin's work. The anthropologist Alexander Goldenweiser soon picked up the issue but stated it so extravagantly as to convert a sound observation into a self-defeating exaggeration, saying that "the meshes of his [Sorokin's] net are spread so wide that all [?] that counts in history slips right through it."[12] And essentially the same point is made by Maquet about differing social and political systems when he notes that "some differences which are significant from a microscopic point of view are neglected. Thus, communism, capitalism, fascism are subsumable under the same category of sensate culture . . . the use of conceptual tools like the three premises of culture will let a rather large number of differences very important in regard to a narrower frame of reference escape."[13]

Inspecting the course of science through Sorokin's macrosociological lens is apt to blur specific developments in science rather than to bring them into sharp focus. For example, in his account of how culture mentalities affect the foci of attention in science—a problem important in its own right—Sorokin observes: " . . . the scientists of Ideational culture would be more *interested* in the study of spriritual, mental, and psychological *phenomena*. . . . Scientists of Sensate culture would probably be more *interested* in the purely material *phenomena* . . ." (italics added).[14] The *comparative* degrees of interest in these two broad classes of phenomena at any one time need not be put in question. But it does divert us from considering the import of the fact that an immense interest has in recent generations developed in the sciences of human behavior which are concerned with "spiritual, mental, and psychological phenomena." Paraphrasing Derek Price's estimate of physical scientists, we have only to remember that more than 90 percent of all social and behavioral scientists that have ever lived are still alive. This great interest in the scientific investigation of man and his works is a historical fact that requires interpretation by the sociology of science, but it is not one readily explained by Sorokin's macrosociological conceptions.

In emphasizing this general point, we should prefer not to be misunderstood. It is not being said that Sorokin's macrosociology of science is *theoretically incompatible* with the more detailed analysis of varying developments of thought and science within each of his major types of culture. It is not a matter of theoretical inconsistency but, rather, a matter of the kinds of inquiry in the sociology of science that tend to be emphasized and those that tend to be neglected in the macrosociological perspective. That is what we mean by saying that, in this respect, Sorokin's theory is a first approximation. It can be and, we argue, should be complemented by intensive inquiries into the connections between types of scientific work by men variously located in the social structure of a particular society.

Much the same issue is involved in Sorokin's treatment of long-run and short-run fluctuations in the modes of thought prevailing in one or another sphere of culture. Sorokin is of course primarily interested in the long-run fluctuations of culture mentalities, which he regards as fundamental to all the rest. But he does attend—for example, in chapter 12 of the second volume of the *Dynamics*—to short-run changes in such scientific theories as atomism, vitalism, and mechanism in biology, abiogenesis, and corpuscular and wave theories of light, going on to note that "across the ever-recurring alternation of these theories, short-time fluctuations may also be perceived. . . ."[15] But these short-run variations do not engage Sorokin's interest; he makes no effort to investigate their social and cultural sources. More specifically, he observes, "The situation in regard to mechanistic and vitalistic conceptions in the present century appears to be one of armed conflict. Both conceptions seem to be existing side by side and both seem to be flourishing."[16] Again, Sorokin does not consider it part of his theoretical commitment to examine the social and cultural conditions under which these opposed biological theories are found in a state of armed coexistence. Yet this would plainly be a major problem for the microsociology of science.

### 3. Cultural Determinism and the Relative Autonomy of Subsystems

Up to this point we have treated Sorokin's general theoretical position in its more extreme and emphatically reiterated form. This position holds that the three types of "culture mentalities" alone determine the form, substance, and development of knowledge in general and of science in particular. It is compactly expressed, for example, in Sorokin's assertion that "Scientific theory thus is but an opinion made 'creditable' and 'fashionable' by the type of the prevalent culture."[17] That theories in science which are not acknowledged as valid by a substantial part of the community of scientists form no significant part of the science of the time is of course the case. But this is a far cry from concluding that scientific theory is *nothing but* a matter of accreditation and fashion. If this were so, it would negate a principal fact about the history of science, the *accumulation* of certified knowledge, albeit an accumulation that proceeds at uneven rates. Whatever else may be disputed, we can scarcely deny that there exists a greater stock of scientific knowledge today than in the past. There is more here than a mere matter of belief and fashion.

In point of fact, Sorokin does not confine himself to the extreme position that holds the development of science to be wholly determined by the prevailing culture mentality. Instead, he introduces two qualifications, the one emphasized as an integral part of his theory and the other treated casually and only in passing. The first qualification assigns a margin of autonomy or independence to each subsystem in a culture, especially the subsystem of disciplined thought and science; the second briefly acknowledges that a differentiated social structure as well as the dominant culture mentality affects the development of knowledge. Both qualifications, and particularly the first of these, are essential to a sound reading of Sorokin's theory.

Possibly because Sorokin himself so often emphasizes the dependence of everything in a sociocultural system upon its "cultural premises," critics of his work understandably take him to subscribe to a doctrine of rigid cultural determinism. The dominant emphasis overshadows the basic restriction upon this restriction upon this doctrine expressly introduced in the first chapter of his *Dynamics*, where each subsystem of a sociocultural system is seen as having a degree of autonomy or independence. Put most generally,

> The autonomy of any system means . . . the existence of some margin of choice or selection on its part with regard to the infinitely great number of varying external agents and objects which may influence it. It will ingest some of these and not others. . . . [O]ne of the most important "determinators" of the functioning and course of any system lies within the system itself, is inherent in it. In this sense any inwardly integrated system is an autonomous self-regulating, self-directing, or, if one prefers, "equilibrated" unity. . . . This is one of the specific aspects of the larger principle which may be called "immanent self-regulation and self-direction."[18]

*The problem then becomes one of developing a theory adequate to account for the different "margins" of autonomy possessed by various kinds of institutions and other subsystems.* So far as we can see, this is another gap—in our accounting, the third gap—in Sorokin's theory as it now stands. Apart from the roughly ascertainable *fact* that particular institutions have a smaller or greater measure of independence of their social and cultural environments, there seems nothing *in the theory* to help us anticipate how this will turn out for various kinds of institutional spheres in various kinds of sociocultural systems.

With regard to this problem, Sorokin would seem to hold a position formally (not of course substantively) like that adopted by Marx and Engels. In view of Sorokin's well-known opposition to Marxist theory, this statement may at first seem to be implausible, not to say extravagant. Yet when theorists are confronted with the same problem, they not infrequently converge in their formal analysis of it, however much they may differ in their substantive conclusions. And this, it seems to us, is the case with Marx and Sorokin in their treatment of the relative autonomy of institutional spheres within society. Consider only these few parallelisms of formal analysis.

Just as Sorokin in the main makes his culture mentalities the effective determinant of what develops in a sociocultural system, so, of course, Marx makes the "relations of production" the "real foundation" which "determines the general character of the social, political and intellectual processes of life."[19] Substantively, Marx and Sorokin could not be further apart: Marx adopts a "materialistic" position in the sense of the social relations of production largely determining the superstructure of ideas;[20] Sorokin adopts an "idealistic" position in which the underlying premises and cultural mentality largely determine the general character of the society and culture, including its social relations. But both agree on the formal position of positing *primary* social or cultural determinants that nevertheless leave room for some degree of independence in the spheres of thought and knowledge.

Just as Sorokin postulates some measure of autonomy for social and cultural subsystems, so does the alter ego of Marx as, for example, when he attributes a degree of autonomy to law. Thus Engels writes:

> As soon as the new division of labor which creates professional lawyers becomes necessary, another new and *independent* sphere is opened up which, for all its general dependence on production and trade, has its own capacity for reacting upon these spheres as well. In a modern state, law must not only correspond to the general economic position and be its expression, but must also be an expression which is *consistent in itself*, and which does not, owing to inner contradictions, look glaringly inconsistent. And in order to achieve this, the faithful reflection of economic conditions is more and more infringed upon. All the more so the more rarely it happens that a code of law is the blunt unmitigated, unadulterated expression of the domination of a class—this in itself would already offend the "expression of justice."[21] [Italics added]

In the Marxist view, if this is true of law, closely connected with economic processes, it is all the more true of other spheres in the "ideological superstructure." Philosophy, religion, and science are in particular constrained by the preexisting stock of knowledge and belief, and are only indirectly and ultimately influenced by economic factors.[22]

> Political, juridical, philosophical, religious, literary, artistic, etc. development is based on economic development. But all these react upon one another and also upon the economic base. It is not that the economic position is the *cause and alone active*, while everything else only has a passive effect. There is, rather, interaction on the basis of the economic necessity, which *ultimately* always asserts itself.[23] [Italics added]

As we have seen, Sorokin puts all this more generally in his concept of the autonomy of logically or functionally unified systems. And he applies the concept, among other cases, specifically to the institutional sphere of science in these words:

> . . . it is not claimed that all scientific theories show, or must show such a connection [with the underlying cultural premises]; many of them can fluctuate independently of our main variables, within their limited sphere of autonomy and the immediate mental atmosphere of their compartment. . . . Due to the *Principle of*

*Autonomy* of any really integrated system, each of the integrated currents of culture mentality studied should be expected to have some margin of this autonomy. . . . [24]

Thus, just as Marx-Engels regard the *pressure toward internal consistency* within each institutional sphere as a source of its comparative independence of the social relations of production that are the "ultimate" determinant, so Sorokin regards the *integration* of a subsystem as a source of its comparative independence of culture mentality as the "ultimate" determinant. The difference of theory is substantive rather than formal.

This brings us back, then, to the third problem of Sorokin's theory to which we have alluded: How does the theory deal with the comparative degrees of autonomy characteristic of different institutional sybsystems in a society? Is the measure of autonomy the same for them all—for religion and law, for science and philosophy? Or is there a theoretical basis for assuming that the degree of autonomy characteristically differs for these subsystems? To raise the question is one thing; to supply a satisfactory answer is quite another. When we suggest that the Sorokinian theory seems to provide no answer, we consider this, rather, as an identifiable and instructive gap than as an observation that undercuts the basis of the theory. The following loose formulation by Engels provides only a suggested clue to the solution, rather than the solution itself: "The further the particular sphere which we are investigating is removed from the economic sphere and approaches that of pure abstract ideology, the more shall we find it exhibiting accidents [i.e., deviations from "the expected"] in its development, the more will its curve run in zig-zag."[25] This suggestion still leaves open the difficult question of how to find out the "distance" of each institutional sphere from the economic sphere. But if Engels left the problem unresolved, so, too, it seems, does Sorokin.

That this is so is further suggested by Sorokin's passing observations on the independent functions of science in any sociocultural system. After pointing out that a society like that of the United States has a "highly integrated and differentiated system of science," while primitive societies have "little developed" systems of science,[26] he notes: "But in some form science will be found as a system in any culture area, because [note the functional assumption] any social group, as long as it lives, must have and does have a minimum of knowledge of the world that surrounds it, of the phenomena and objects that are important for its survival and existence. No group entirely devoid of any knowledge can exist and survive for any length of time."[27] It is not the functional assumption of some indispensable minimum of authentic knowledge that concerns us here; rather, it is that this assumption ascribes an independent function to "science" in every society, so that we see Sorokin once again implying that science is not merely the reflection of the culture mentality but has its own functional basis as well.

After this extended discussion of the principle of autonomy of subsystems in Sorokin's theory and of the unfilled gap in that theory, we may turn for a moment to the second of his restrictions on the cultural determination of science. This must be brief, not because we consider it unimportant, but because, as we have intimated in the foregoing section, Sorokin has elected to give it only fleeting attention in his own work. This restriction upon the determination of thought by the general culture mentality deals with the connections between the internal differentiation of the social structure and the character of the diversified thought that obtains in the society. It deals, in the language of the foregoing section, with problems in the microsociology of science rather than its macrosociology.

Symptomatically enough, Sorokin touches upon this only in a long footnote. Moreover, the note is not in his *Dynamics* where he most fully develops his sociology of knowledge and science but in his later general introduction to sociology, *Society, Culture and Personality*. Only there, in discussing what he describes as the "non-logicity" of ideas, does Sorokin remark of Mannheim's analysis:

> He rightly looks for the cause in the group affiliations of a person; but . . . his theory remains vague, and in many respects incorrect. Meanwhile the real reasons for non-logicity are at hand. They are the nature of one's group affiliations and one's cultural affiliations. . . . Unfortunately how our social affiliations influence our logic and judgments is still but little known. The so-called "sociology of knowledge" has hardly reached a clear formulation of this problem.[28]

In short, Sorokin here acknowledges the saliency of the problem of how, within the same culture, differences in social status and group affiliations affect the nature of nonlogical sentiments, of logical thought, and, presumably, of scientific inquiry.  No good purpose would be served in discussing Sorokin's appraisal of the current state of the sociology of knowledge, particularly since we are agreed that singularly little empirical inquiry has been developed in this field of enduring intellectual interest. What is pertinent is not so much the appraisal of work left undone, but the *theoretical* issue that such inquiry, advocated by Sorokin, presupposes the probability of distinct lines of thought that will differ according to the social status and group affiliations of men of science and of the intellect generally. For this implies a conception of the sociology of knowledge that allows for significant variability in the ideas and knowledge developed *within* a particular culture—variability that results from social differentiation—and so supports our interpretation that Sorokin's macrosociology of knowledge does not deny in principle the pertinence of socially differentiated sources of knowledge. This, then, indicates once again the theoretical receptivity on the part of Sorokin—although he has chosen not to pursue this tack for himself—to the notion that the broad cultural mentality does not fully determine the character of knowledge but allows for significant and socially patterned variations in that knowledge.

Thus, if we take account of Sorokin's two basic qualifications to the determination of science and other knowledge by the prevailing culture mentality, we find that his theoretical position is not as far removed as it would seem from that adopted by other sociologists of science. His theory sets us the empirical task of trying to ferret out the ways in which culture and social structure affect the development of knowledge, allowing some measure of independence to the requirements internal to each branch of knowledge and science. The appearance of this volume of papers affords an opportunity for Sorokin to set out his present thinking on this question central to the sociology of knowledge. It will then be possible to decide whether Maquet is correct, or merely vague, in his conclusion that "For Sorokin, it is certain that the existential factor [i.e., the social structure] is the least important. . . . He considers that the premises of culture are really the most important factors [*sic*] for the determination of mental productions. . . . We can say that in reality [for Sorokin] the cultural premise truly exercises a predominant influence."[29] Whatever else can legitimately be said of Sorokin's theory, it cannot be described as a theory of "factors," of great, middling, or slight "importance." When Sorokin undertakes to translate Maquet's fuzzy expressions—such as "least important" and "most important factors"—into ideas that are definite enough to bear inspection, we shall be the better able to appreciate his current position on the place, in his theory, of the socially patterned distribution of types of knowledge that is found within each kind of culture.

## 4. Empirical Research: Quantitative Indicators in the Sociology of Science

Thus far we have attended to certain components of Sorokin's macrosociological theory of science, singling out those that give rise to theoretical issues that would profit from further clarification. In doing so, we have raised three questions about puzzles that persist in Sorokin's theory: first, how the theory escapes from an emanationist position, which postulates underlying culture mentalities that seem to include, in their definition, what is later said to be an expression of these mentalities; second, and to our mind basically, how the theory accounts for the socially patterned distributions, within a particular type of culture, of diverse modes of thought that do not correspond to the prevailing tendencies; third and correlatively, how the theory deals with the comparative degrees of autonomy characteristic of various subsystems within a sociocultural system, so that it can treat the extent of observed autonomy not simply as an empirical given but as theoretically explainable.

But since Sorokin's is an empirically connected theory, rather than one presented as a set of abstractions remote from systematically assembled data, we have now to turn to selected aspects of his empirical inquiries in the sociology of science. And here the most striking feature of Sorokin's work is the creation of massive accumulations of social and cultural statistics, designed to serve as basic empirical indicators of underlying changes in the rate and character of social and cultural changes.    Surely more than any other single scholar dealing with problems in the sociology of knowledge, Sorokin has in effect heeded the maxim of the French social historian Georges Lefebvre, "Il faut compter."

As one of us has had occasion to note before:

> Studies in historical sociology have only begun to quarry the rich ore available in comprehensive collections of biographies and other historical evidence. Although statistical analysis of such materials cannot stand in place of detailed qualitative analysis of the historical evidence, they afford a *systematic* basis for new findings and, often, for correction of received assumptions. . . . The most extensive use of such statistical analysis is found in Sorokin's *Dynamics*.[30]

When we speak of Sorokin's "creation" of these social and cultural statistics, we do so advisedly. For, unlike the operations of governmental bureaus of the census, there are few kinds of social bookkeeping that systematically record evidence on the kinds of intellectual developments with which Sorokin's theory requires him to concern himself. (Statistics of patents for inventions in the modern period and data on the numbers of books published in various fields practically exhaust all that is readily available on these subjects.) And so, in spite of Sorokin's remarkably ambivalent attitude toward the use of sociological statistics, he found himself required, by the implications of his own theory, to assemble statistics that would testify to the degree of integration empirically found in each of his theoretically constructed types of culture.

By assembling these statistics, Sorokin boldly confronted the problem of how to find out the *extent* to which cultures are in fact integrated. Despite his vitriolic comments on the statisticians of our sensate age, he recognized that to deal with the extent of integration implies some statistical measure. Accordingly, for the field of knowledge, he developed numerical indices of writings and authors in each time and place, had these coded and classified in appropriate categories, and thus assessed the comparative frequency and inferred influence of various systems of thought.

In the sociology of science, for example, the data cover the period from 3500 B.C. to the twentieth century, being based upon counts from such standard sources as Darmstädter's *Handbuch zur Geschichte der Naturwissenschaften und der Technik*, F. H. Garrison's *Introduction to the History of*

*Medicine*, and the ninth edition of the *Encyclopaedia Britannica*. Counts such as these provide the basis for empirical confirmation of the theoretically derived proposition that "The rate of scientific development tends to become slow, stationary, even regressive in Ideational cultures ... becoming rapid and growing apace in Sensate cultures...."[31]

There is neither need nor space to report the limitations of his quantitative indicators as these are set out by Sorokin.[32] In any event, he concludes that, whatever their limitations, the indicators provide a valid and reliable measure of fluctuations in the rate of scientific discovery and technological invention as well as of other intellectual and artistic expressions of the culture. That is why he is prepared to assert that "Not only do the first principles and categories of human thought fluctuate, but also most of the scientific theories of a more or less general nature."[33] Plainly, he bases his empirical conclusions very largely upon these cultural statistics.

In view of the basic part played by these statistics in his sociology of knowledge, Sorokin adopts a curiously ambivalent attitude toward them. This can be seen in his approval of the remark by Robert E. Park that his statistics are merely a concession to the prevailing sensate mentality and that "if they want 'em, let 'em have 'em."[34] Park's facetious remark need not be allowed to obscure the symptomatic nature of Sorokin's response. It is indicative, we believe, of Sorokin's fundamental ambivalence toward criteria of scientific validity, an ambivalence deriving from his effort to cope with quite disparate "systems of truth." In view of the vast effort that went into compiling the cultural statistics that underlie Sorokin's work in the sociology of science, it would not only be facetious but thoroughly irresponsible to conclude that these systematic data were merely trappings considered necessary to "convince the vulgar." The fact is that Sorokin's empirical descriptions are very largely based on these statistics. They are essential to his argument. To remove them would not be to remove a facade, leaving the essential structure of his theory intact; it would be to undercut his macrosociological theory of science and to leave it suspended in the thin air of unrestrained speculation.

This, then, leads to another, the fourth, question and this one two-pronged, which the dialogue of this book may help answer. In view of Sorokin's ambivalence toward social and cultural statistics, about which we shall have more to say, we must ask: What is his current and perhaps consolidated position with regard to the place of such statistics in sociological inquiry, primarily in the sociology of science and, by implication, in other branches of sociology as well? Further, how does his discussion of this question help clarify his position on the criteria of scientific truth, which he adopts: does he regard systematic evidence of the kind caught up in his statistics as merely a mode of communication to his scientific compeers in a sensate culture or as a substantial basis both for confirming and for developing his theory?

It is these cultural statistics, moreover, that serve to highlight once again two of the principal questions that we consider still unresolved in Sorokin's sociology of knowledge: the question of how the theory accounts for observed variations in the modes of knowledge within a culture and the question of accounting for the distribution of these differences among various groups and strata in the social structure. Take just one case in point. Sorokin describes empiricism as "the typically sensate system of truth." The last five centuries, and more particularly the last century, represent "sensate culture *par excellence!*"[35] Yet even in this flood tide of sensate culture, Sorokin's statistical indices show only some 53 percent of influential writings to be characterized by "empiricism." Furthermore, in the earlier centuries of this sensate culture, from the late sixteenth to the mid-eighteenth, the indices of empiricism are consistently *lower* than the indices for rationalism (which, in the theory, is associated with an idealistic rather than a sensate culture). The statistical indicators, then, show that the notion of a "prevailing" system of truth needs to be greatly qualified, if it is to cover both the situations in which it represents a bare statistical "majority" and even a statistically indicated minority in the writings of a period.

Even more is implied by Sorokin's statistics. For the main purpose of our observations is not to raise the question of the extent to which Sorokin's conclusions coincide with his statistical data: it is not to ask why the sixteenth and seventeenth centuries are said to have a predominantly "sensate system of truth" in the light of these data. Rather, the purpose is to suggest that, even on Sorokin's own premises, the general characterizations of historical cultures as sensate, idealistic, or ideational constitute only a first step in the analysis, a step which must be followed by further detailed analyses of deviations from the central tendencies of the culture. Once Sorokin has properly introduced the notion of the *extent* to which historical cultures are in fact integrated, he cannot, in all theoretical conscience, treat the existence of types of knowledge which differ from the dominant tendencies as evidence of a mere "congeries" or as a merely accidental fact. It is as much a problem of the sociology of science to account for these substantial "deviations" from the central tendency as to account for these tendencies themselves. And for this, we suggest yet again, it is necessary to develop a theory of the socio-structural bases of thought in a fashion that a cultural-emanationist theory does not permit.

Apart from these theoretical implications, Sorokin's statistics presented in his *Dynamics* afford an occasion for exploring further the intellectual grounds of his ambivalence toward social and cultural statistics altogether. As is well known, Sorokin devotes a considerable portion of his book *Fads and Foibles in Modern Sociology* to an attack on "quantophrenia," or an uncritical devotion to faulty statistics. That quantitative methods in sociology can be, and have been, abused is surely not in question, any more than that qualitative methods, based on ill-devised and ill-confirmed impressions, can be and have been abused. And surely, no sober man will declare himself in favor of faulty craftsmanship, unsound assumptions, and mistaken inferences. The question is therefore not one of identifying this or that case of a fallacy in quantitative analysis in sociology but, rather, one of setting out the criteria and limits of sound quantitative analysis. And since so much of Sorokin's work in the sociology of science is pervaded by empirically grounded statistics, this question becomes thoroughly germane to our discussion.

What, then, are Sorokin's criteria for the appropriate use of social and cultural statistics?[36] We find it decidedly easier to raise the question than to answer it. Indeed, we raise the question in the hope that Sorokin will give his pointed and definite answer to it. This becomes all the more pertinent when we find that some scholars are prepared to adopt an even more extreme perspective on social and cultural statistics than Sorokin's own. Werner Stark, for example, says of Sorokin's *Dynamics* that

> our criticism . . . is one of principle. His whole procedure assumes *a radice* the possibility of quantifying what is qualitative, and this is almost like supposing it is possible to square the circle. A book, or a work of art, is all quality [n. b.], because it is all spirit. . . . It is to be feared that the sociology of knowledge will never be able to get much assistance from statistical techniques. Much as we may regret the fact, it will always have to rely heavily on the more cumbersome monographic and descriptive methods.[37]

The issue is even more stark than Stark apparently supposes. For everyone in his senses would agree that what is "inherently" and "exclusively qualitative" cannot, by definition, be quantified. It is really asking too much to ask us to reject a strict tautology. But when we get down to cases, the crucial question, of course, is begged by such an affirmation; the question is precisely one of establishing criteria of what is irrefragably qualitative and of what, in some aspect and degree, can be reasonably and usefully quantified. And since, in our opinion, Sorokin has wisely and justifiably counted *aspects* of complex works of science and art, it would be helpful to have him clarify the sense in which he found these to be quantifiable.

Sorokin's restatement of his position on the issue of such quantification would be particularly instructive in view of what he has said about the issue in his *Fads and Foibles*. At one place in that work, for example, he declares that

> only through direct empathy, co-living and intuition of the psychosocial states can one grasp the essential nature and differences . . . of religious, scientific, aesthetic, ethical, legal, economic, technological, and other cultural value-systems and their subsystems. Without the direct living experience of these cultural values, they will remain *terra incognita* for our outside observer and statistical analyst. . . . These

methods are useless in understanding the nature and difference between, say, Plato's and Kant's systems of philosophy, between the ethics of the Sermon on the Mount and the ethics of hate, between Euclidean and Lobachevskian geometry and between different systems of ideas generally. Only after successfully accomplishing the mysterious inner act of "understanding" each system of ideas or values, can one classify them into adequate classes, putting into one class all the identical ideas, and putting into different classes different ideas or values. Only after that, can one count them, if they are countable, and perform other operations of a mathematical or statistical nature, if they are possible. Otherwise, all observations and statistical operations are doomed to be meaningless, fruitless, and fallacious simulacra of real knowledge.[38]

It would no doubt be generally agreed that a proper understanding of cultural content is required for it to be validly classified so that specimens in each class can then be counted. The vast compilations and counts of such data in the *Dynamics testify* that Sorokin also thinks this can be done. But is it not too stringent a criterion to require a "direct living experience of the cultural values" in order for them to be classified and counted? Some substantial knowledge about the materials in hand is of course necessary but this would seem to fall far short of the extreme requirement exacted by Sorokin. We cannot assume that all of Sorokin's research associates and assistants had a "direct living experience" of the many thousands of scientific discoveries, technological inventions, philosophical doctrines, and art objects that they classified and counted in order to provide an empirical test of Sorokin's ideas. It is certain that one of his research assistants, R. K. Merton, had no such demanding experience of the almost 13,000 discoveries and inventions he computed on the basis of the Darmstädter *Handbuch*, just as it is probable that J. W. Boldyreff, another of his assistants, had no such experience of the thousands of scholars, scientists, artists, statesmen, etc., mentioned in the ninth edition of the *Encyclopaedia Britannica*, who were classified and assigned weights on the basis of the amount of space devoted to them in the *Encyclopaedia*.

Nevertheless, there is internal evidence that these counts were not vitiated by limited knowledge (though, we suggest, knowledge enough for the purpose in hand). For independent classifications and counts of different but theoretically related materials produced much the same empirical results. As Sorokin reports for one such case dealing with data on the "empirical system of truth (of senses)" and data on the rate of scientific discovery:

The items and the sources were entirely different and the computations were made by different persons who were not aware of the work of the other computers. (Professors Lossky and Lapshin had no knowledge of my study, and Dr. Merton, who made the computation of the scientific discoveries, was unaware not only of my study but also of the computations made by Professors Lossky and Lapshin.) Under the circumstances, the agreement between the curve of the scientific discoveries and inventions and the curve of the fluctuations of the influence of the system of truth of senses is particularly strong evidence that the results obtained in both cases are neither incidental nor misleading.[39]

In a word, the quantification of cultural contents cannot, need not, and is not intended to reproduce the entire complex whole of each item entering into the computation. Only selected aspects and attributes are classified and counted. And for this purpose, full, detailed, and empathic understanding of each cultural item is not, apparently, required. It would therefore be instructive to have Sorokin redirect his attention to the seeming discrepancy between the actual practice employed in quantifying cultural items in the *Dynamics* and the far more demanding criteria for such quantification proposed in the *Fads and Foibles*. What Sorokin actually *does* in the one case seems to us more compelling than what he *says* in the other. In making this observation, we only adopt and adapt the sage advice of Albert Einstein: "If you want to find out anything from the theoretical physicists about the methods they use, I advise you to stick closely to one principle: don't listen to their words, fix your attention on their deeds."[40]

All this allows us to note that not the least advance in sociology during the last century or so is reflected in the growing recognition that even crude quantitative data can serve the intellectual purpose of enabling the sociologist to reject or to modify his initial hypotheses when they are in fact defective. To see this change in outlook we have only to contrast the encyclopedic effects of a Comte with those of a Sorokin. Comte handles scattered facts gingerly and infrequently, as though they were unfamiliar and even dangerous things; he does not think of so assembling systematic arrays of data that they could, in principle, put his intuitive or reasoned guesses to the test of empirical reality. Sorokin drenches us in quantitative facts—for example, in the *Dynamics*, but not only there—and thus provides both himself and his readers with the occasion for matching theoretical expectations and empirical data. This practice would seem particularly required when scholars turn to the sociological drama of large-scale changes in the cultures and social structures that make up the framework of world history. For entirely qualitative claims to facts prove to be excessively pliable, easily bent to fit the requirements of a comprehensive theory. But if it is to be more than a dogma, a theory must state the empirical observations that will be taken to disprove it or, at least, to require its substantial revision. Independently collected, systematic and quantitative data supply the most demanding test called for by such an empirically connected theory. And that Sorokin also thinks this to be the case seems implied by the way in which he has gone about his task of conducting empirical inquiries in the sociology of science.

## 5. Relativism and the Criteria of Scientific Truth

We have alluded, once or twice, to the problem confronted by Sorokin of locating his own work in one or another of the "systems of truth," which he makes distinctive of each of his three major types of culture. What criteria of truth does he employ in setting about his own work? Is he a thoroughgoing relativist, regarding scientific truth as *nothing but* a matter of satisfying the different criteria that obtain in each particular type of culture? Does he consider each system of truth just as compelling (or as arbitrary) as the next? Does he see himself as a creature of contemporary sensate culture, subject to contemporary criteria of scientific truth, or has he found an Archimedean point to stand upon, which enables him to move beyond these criteria? If so, what is this point and how does he assure himself and his prevalently sensate readers that it is an effective and justifiable one? In short, how does Sorokin try to escape the relativistic impasse?

This barrage of questions—at bottom, they of course comprise only one question—is something more than a matter of rhetoric. We are genuinely puzzled and unable to identify, with any assurance, the position taken by Sorokin on this matter. Our confusion is further confounded by what seems to be Sorokin's indecisive and possibly changing conception of science in today's sensate culture. We recall his statement that "In a Sensate society and culture, the Sensate system of truth based upon the testimony of the organs of senses has to be dominant."[41] But, it turns out, this statement is only a gross approximation. For reason enters into the system as well. The sensate method of validation requires "Mainly the reference to the testimony of the organs of senses. . . , supplemented by logical reasoning, especially in the form of mathematical reasoning. But even the well-reasoned theory remains in the stage of pure hypothesis, unproved until it is tested by the sensory facts; and it is unhesitatingly rejected if these 'facts' contradict it."[42] And, for our immediate purposes, finally, he writes that the sensate system of scientific truth "possesses some of the elements of the rationalistic system of truth in various forms; in the forms of the laws of logic which are obligatory for scientists and which are hardly mere results of the sensory experience; in that of deductions, which are incorporated in the queen of these sciences, mathematics; of many conceptual elements in the form of the fundamental concepts and principles of the sciences; and in several other forms."[43]

With this statement Sorokin seems to have returned, from a distant point of departure, close to the position which, except for turns of language, is that generally adopted by working scientists in our time. Intuition, hunch, and guess may, and often do, originate ideas, but they do not provide a sufficient basis for choosing among ideas. Logical analysis and abstract reasoning interlock with empirical inquiry and it is only when the results of these two prove consistent that contemporary scientists consider them to be an authentic part of validated scientific knowledge. However much Sorokin may on occasion seem to take joy in the system of truth described as characteristic of an idealistic culture, he nevertheless *practices* under the rules of a sensate system. That, we suppose, is what lies behind his footnoted remark: " . . . however surprised a contemporary partisan of scientism may be at my impartiality in 'observing and ascribing' the existence of various systems of truth. . . , he has to countenance it because they are empirical facts witnessed by the testimony of our organs of senses, as will be demonstrated further. In other words, in my study I shall intentionally follow the 'empirical system of truth' which must be convincing to such a partisan of 'scientism.' "[44]

Here Sorokin says that he adopts as criteria for his own work that complex of rational discourse and empirical data which is characteristic of a sensate science. But he implies that he does so only as a *façon de parler*. Yet this reply-in-advance to our question seems facile rather than adequate. Does it mean that Sorokin as a social scientist is truly prepared to abandon empirical tests of his ideas? that he is ready to propose characterizations of historical societies and cultures which are at odds with the empirical evidence he has assembled? We suspect not. The composite of reason and ordered experience seems to us precisely what Sorokin in fact employs as a guide to his own inquiry and as a measure of the acceptability of the results of the inquiry. Intuition, scriptures, chance experiences, dreams, or whatever may be the psychological source of an idea. (Remember only Kekulé's dream and intuited imagery of the benzene ring, which converted the idea of the mere number of atoms in a molecule into the structural idea of their being arranged in a pattern resulting from the valences of different kinds of atoms.) But whatever the source, the idea itself must be explored in terms of its implications and these implications then examined in terms of how far they hold empirically.

To put the issue directly and so to afford Sorokin an occasion in this dialogue for further clarifying his position, we suggest that, whatever asides may be tucked away in footnotes, Sorokin adopts, in the course of his inquiries in the sociology of science, a thoroughgoing commitment to the combined criteria of internal consistency and empirical observation that are the mark of scientific work in our sensate age.

Sorokin's image of sensate science notwithstanding, the fact is that concepts and rules of reasoning are no mere props in modern science. They are as indispensable as the testimony of the senses. We call only one witness, although many more are waiting in the corridors of today's science:

Our experience hitherto justifies us in believing that nature is the realization of the simplest conceivable mathematical ideas. I am convinced that we can discover by means of purely mathematical constructions the concepts and the laws connecting

them with each other, which furnish the key to the understanding of natural phenomena. Experience may suggest the appropriate mathematical concepts, but they
most certainly cannot be deduced from it. Experience remains, of course, the sole
criterion of the physical utility of a mathematical construction. But the creative
principle resides in mathematics. In a certain sense, therefore, I hold it true that
pure thought can grasp reality, as the ancients dreamed.[45]

This is not the voice of the thirteenth-century Robert Grosseteste speaking;
it is the voice of the decidedly twentieth-century Albert Einstein.[45]

Moreover, as the history of science during the last centuries testifies, not
only can empirical data challenge established concepts and theories, but concepts and theories often challenge the superficial testimony of the senses. It is
a familiar part of everyday practice in science to reject misleading empirical
impressions when these run counter to theories that have themselves been
firmly embedded in scientific thought. Any sharp separation of reason and
empirical data in contemporary science must therefore distort much of the
operative reality. Work in the scientific laboratory rests upon both, with one
or the other raising questions that must be resolved by a congruence between
them. Only then is there a reasonable prospect that an idea or a finding will
enter permanently into the repertory of science. And this sensate conception
of science, we suggest, is basic to Sorokin's own work, his incidental dis-
claimers notwithstanding. This, at least, is a sixth puzzle, which Sorokin
might helpfully unravel in his part of the dialogue.

## 6. The Cumulation of Scientific Knowledge

The issue we have just identified leads us directly to still another question
about Sorokin's theory of social and cultural dynamics, this one cutting
deeply enough to isolate, for a moment, his sociology of science from the rest
of his theory. The fact that it is an issue hoary with age does not make it any
the less in point. We refer, as the heading of this section implies, to the particular sense in which science, as distinct from other spheres of culture, tends to
be accumulative. In our view, this raises a question, deeply imbedded in
Sorokin's theory of culture change, that requires him to consolidate his role as
sociological historian and as sociological theorist.

As sociological theorist, and on his own accounting, Sorokin has identified
two full cycles of ideational-idealistic-and-sensate phases in Greco-Roman
and Western cultures. He sees a third sensate phase beginning roughly in the
fifteenth or early sixteenth century. In his vocabulary of abstract types of
culture, one ideational phase in history is much like the other; one idealistic
phase is much like the next; and one sensate phase is much like the rest. For
these are described and analyzed in terms of general categories and criteria,
in the light of which they seem to be "of the same kind."

As sociological historian, however, Sorokin must reckon with quite another question. Whatever his theory may identify as similar *kinds* of culture phases, there remains the historical question of the extent to which prior cultural products accumulate and become the possession of men living in a later period of the same or differing cultural type. The cycles of cultural change do not start anew. Particularly with regard to science, each succeeding historical phase makes use of antecedent knowledge on which it builds. In this more nearly concrete, historical sense, the sensate phase of the last centuries is *not*, of course, identical with the sensate phase of the preceding cycles. The phases are *alike*, in terms of the abstract categories employed by Sorokin, else they would not be classified as sensate. But they differ—and science remains our test case of this—in that some of the cultural products of the past are available to those lviing in the later phase. Science did not start anew in the sensate phase, said to begin early in the sixteenth century; as historians of science periodically remind us, it built upon the selective accumulation of what had gone before.

All this seems evident enough. Yet, possibly because Sorokin is adamantine in his rejection of a unilinear doctrine of cultural change, he tends to neglect the *implications* of selective cultural accumulation[46] for his theory. It is this accumulation and its consequences that distinguish the sensate culture of the twentieth century from the sensate culture of, say, the Hellenistic period. To describe both periods as sensate is justified only abstractly but not historically. For the accumulation of scientific and technological knowledge makes a difference that can make a very great difference to men living in the later of these sensate phases. To say that the thesis of a unilinear accumulation of knowledge cannot qualify as historical truth is one thing. But to ignore selective accumulation of knowledge is quite another. (To put it vulgarly, the Hellenistic Greeks did not have a body of knowledge about quantum mechanics or a technology of spacecraft; or, to scramble legend and history, Icarus really cannot be equated with the astronauts Gargarin and Glenn.)

What we have been saying raises two related questions about Sorokin's macrosociology of science. These are questions about what he is prepared to take as significant similarities and what as significant differences in the scientific knowledge found in historical eras of the same abstract type.

The first question comes to the fore when we examine again his observations on short-run fluctuations of particular scientific theories in various periods.[47] He summarizes his judgment in these words: " . . . as far as mere oscillation is concerned, there probably has been no scientific theory which has not undergone it, and, like a fashion, now has been heralded as the last word of science, and now has fallen into disrepute."[48] This judgment leads us to ask in what sense recurrent sets of ideas constitute one and the same theory that now finds general acceptance and later, rejection, only to be accepted again, still later. To consider one of the instances cited by Sorokin as a case of fluctuations in a theory, in what sense is present-day "atomistic theory" to be taken as the same as "atomistic theory" in ancient Greece? Similarities are there, of course, but also, obvious and significant differences. And it is these accumulative differences in what is on the surface the same kind of scientific theory that constitute an advance in science. To attend only to the formal similarity is to jettison the historically significant differences that enable present-day atomic theory to deal with problems in science that could not, of course, even be dreamt of by the Greeks. Or take the case of fluctuations in the long history of "the theory" of biological evolution, which has been so often traced. Darwin's was not just another version of evolution; it differed from what had gone before by beginning to specify the processes through which the evolution of species took place. Again, to identify Darwin's or later evolutionary theory with ancient versions is to ignore that aspect of the development of science that leads to an enlarged scientific knowledge: selective accumulation. To attend only to similarities between early and later versions of a theory is to become subject to adumbrationism, "the practice of claiming to find dim anticipations of current scientific discoveries in older, and preferably, ancient work by the expedient of excessively liberal interpretations of what is being said now and of what was said then."[49] This is a practice that can only stir up anew the obsolete quarrel between the ancients and the moderns.

The second question raised by Sorokin's relative neglect of the selective accumulation of scientific knowledge has to do with his diagnosis of the present condition of science and his forecast for its immediate future. Knowing Sorokin's sentiments about sensate culture, we can anticipate that his picture of the present state of science will be a gloomy one, and he does not disappoint us:

> One can turn to any field of science now and find first of all a multitude of different theories and sometimes even opposite hypotheses fighting one another for "recognition" as true theory. Such an opulence of contradictions and mutual criticism does not permit any certitude, especially concerning the most important principles, and therefore fosters more and more uncertainty. . . . If such a situation continues—and empiricism, as long as it is dominant, cannot help continuing it—the incertitude will increase. . . . The boundary lines between knowledge and nonknowledge thus are bound to become less and less clear. . . . In such circumstances the truth of senses can easily give way to a truth of faith. In other words, neither doubt, nor uncertainty, nor changeability of the scientific theories can be pushed too far without destroying science itself and its truth. Contemporary science has already possibly gone too far in that direction and therefore is already exposed to danger.[50]

This is strong language. It is the prophetic utterance of a sociological Jeremiah. But, on *that* account, it is not to be lightly dismissed. No one who sat in Sorokin's classes during the 1930s is apt to forget his annual impassioned lecture announcing that one day men of science would create the possibility of destroying all that lives on the earth and that when that day comes, some of these men will be curious to see what really happens when the button is pressed.[51]

From this apocalyptic vision, we return to Sorokin's forecast of a decline in science. This raises again the question of how his theory takes account of the cross-cultural accumulation of scientific knowledge. When Western society largely turned its back on science—in Sorokin's overview, from the third to the eleventh centuries—it turned from a comparatively small stock of accumulated knowledge.[52] It is vastly different in our own sensate age. We are the legatees and initiators of an incomparably greater body of scientific knowledge and of an associated technology not so easily put to one side. To base a prediction on the two preceding cycles of culture would seem hazardous at best; to predict the decline of science in our world means to discount the immensely greater store of science that has accumulated since the last sensate phase and so to treat it as though it were of a piece with the limited scientific knowledge of ancient Rome.

Nor is it evident that confidence in science as a source of knowledge shows signs of diminishing. True, we find expressions of hostility toward science, largely because of the social consequences of some of the technology it has made possible. Science is seen as originating those engines of human destruction that may plunge our civilization into everlasting night and confusion. But there is little of that alienation from science that Sorokin believes to be immanent in the very development of contemporary science. The tonicity of scientists themselves seems more aptly expressed by C. P. Snow when, speaking of what happened in science during two decades at Cambridge University, he says: "I was privileged to have a ringside view of one of the most wonderful creative periods in all physics."[53] He then goes on to describe "a much louder voice, that of another archetypal figure, Rutherford, trumpeting: 'This is the heroic age of science! This is the Elizabethan age!' "[54] And finally, he expresses his conviction that this is a revolutionary time for science, a time in which to take joy in science:

> About two years ago, one of the most astonishing experiments in the whole history of science was brought off. . . . I mean the experiment at Columbia by Yang and Lee. It is an experiment of the greatest beauty and originality, but the result is so startling that one forgets how beautiful the experiment is. It makes us think once again about some of the fundamentals of the physical world. Intuition, common sense—they are neatly stood on their heads. The result is usually known as the contradiction of parity.[55]

That each such advance in science enlarges our awareness of how little is still known is a judgment that has been endlessly reiterated by scientists, particularly the greatest among them.[56] But this does seem far removed from the portrait of uncertainty and confusion among scientists painted by Sorokin. In any case, it would be instructive to have him restate the place occupied by the fact of the accumulation of scientific knowledge in his macrosociological theory of science.

## 7. Themes of the Dialogue

And here we must stop putting questions in detail. Not that other questions fail to make their appearance as we continue to study Sorokin's macrosociology of science. We encounter the question, for example, of Sorokin's appraisal of the current condition of social science. Is it in a thoroughly parlous state, as he suggests in his *Dynamics*,[57] or is it, as he suggests later in *Society, Culture and Personality*, "entering the stage of a new synthesis and a further clarification of its logical structure"?[58] Or again, we meet the question: How does Sorokin see the relations between science and other social institutions, in particular, the institution of religion? Are the relations between the two confined, as he suggests, to those of active combat or of absorption of one by the other, with "rarely, if ever, close cooperation between them"? The matter appears more complex than that. At least, the work of Alfred North Whitehead and others finds that some religions have inadvertently lent support to the pursuit of science and that, apart from the times of conspicuous conflict between them, the institutions of science and religion have not infrequently been mutually supporting.

And so we might continue to raise further questions for Sorokin to reexamine. But like those we have just mentioned, these would touch only the surface of Sorokin's sociological theory of science. It might therefore be more useful to wind up our discussion by recapitulating the puzzles and questions that seem to us unresolved in that theory. This would afford Professor Sorokin an occasion to give us the benefit of his current thinking on these issues and even, perhaps, to divest himself of ideas which once had a definite place in his evolving theory but which now, in the light of further inquiry and reflection, he no longer sees any need for retaining.

In short summary, then, these are the major questions that puzzled us as we reworked our way through Sorokin's sociology of science.

1. Does the theory really adopt an emanationist position, which assumes that the principal features of science and knowledge in a particular culture merely emanate from the culture mentality that underlies it? And since the culture mentality seems to include, in its definition, what are later said to be expressions of that mentality, must we not take this as, rather, an implied definition than an empirically testable hypothesis?

2. What is there in the theory to account for the variability of thought and science *within* each of the societies and cultures that are generally character-ized as being of one or another type: as ideational, idealistic, or sensate?

3. Since integrated subsystems in a culture are said to have a margin of autonomy, a degree of independence of their social and cultural environment, how does the theory account for the margins of autonomy exhibited by the various subsystems? By way of example, does the theory lead us to expect the same or differing margins of autonomy for science and politics, for religion, law, and the economy?

4. What place is assigned, in this theory, to the connections between social differentiation and knowledge? How does the theory deal with the possibility that differences in the social location of men of knowledge, in their statuses and group affiliations, affect the character of what they take as authentic knowledge and what they produce as new claims to knowledge?

5. In view of his variously expressed ambivalence toward social and cul-tural statistics, what is Sorokin's current position on the place of such statis-tics in the sociology of science and, by implication, in other branches of sociology as well? Does he consider cultural statistics of the kind employed in the *Dynamics* as only a means of communicating with his sensate compeers or as also a basis for testing and developing his theory? And since the very stringent criteria of sound quantitative analysis he advocates in *Fads and Foi-bles* do not seem fully met even by his own cultural statistics in the *Dynamics*, would we do better to take his precepts or his practice as guidelines to quanti-tative inquiries in sociology?

6. Which criteria of scientific truth are utilized in Sorokin's own theory? How does it escape from the relativistic impasse of making scientific truth only a matter of taste, in which each type of culture prescribes its own crite-ria? Does Sorokin consider each system of truth just as compelling or just as arbitrary as the next?

7. How does the theory take account of the accumulation of scientific knowledge? Does this accumulation make our sensate period different from the sensate cultures that have gone before?

If we are right in supposing that the foregoing questions direct us to gaps in this macrosociological theory of science, then perhaps the gaps will be bridged in future dialogue.

## Notes

1. Robert K. Merton, *Social Theory and Social Structure* (rev. ed.; Glencoe, Ill.: Free Press, 1957), p. 466.
2. Jacques Maquet, *The Sociology of Knowledge* (Boston: Beacon Press, 1951), pp. 135, 187.
3. W. Stark, *The Sociology of Knowledge* (Glencoe, Ill.: Free Press, 1958), p. 226.

4. Sorokin, *Social and Cultural Dynamics*, vols. 1–3 (New York: American Book Co., 1937); vol. 4 (New York: A.B.C., 1941), 1:72–73.

5. Ibid., vol. 1, p. 67.

6. Ibid., vol. 2, p. 5.

7. Ibid., vol. 1, p. 73.

8. Maquet, *Sociology of Knowledge*, p. 135.

9. For a comparison of the macrosociology and microsociology of knowledge, see Stark, *Sociology of Knowledge* pp. 19–37; also Maquet, *Sociology of Knowledge*, passim.

10. So far as we can see, on only one occasion does Sorokin relate the internal differentiation of a society to any aspect of the types of thought obtaining in that society. This he does tangentially when he contrasts the tendency of the "clergy and religious landed aristocracy to become the leading and organizing classes in the Ideational, and the capitalistic bourgeoisie, intelligentsia, professionals, and secular officials in the Senate culture. . . ." *Dynamics*, vol. 3, p. 250. See also his account of the diffusion of culture among the social classes, vol. 4, pp. 221ff.

11. Cf. Merton, *Social Theory* pp. 466–67.

12. "Sociologos," *Journal of Social Philosophy*, July 1938, p. 353, cited in Sorokin, *Dynamics*, vol. 4, p. 291n.

13. Maquet, *Sociology of Knowledge*, p. 199.

14. *Dynamics*, vol. 2, p. 13.

15. Ibid., vol. 2. p. 446.

16. Ibid., vol. 2, p. 454.

17. Ibid., vol. 2, p. 455.

18. Ibid., vol. 1, pp. 50–51.

19. Karl Marx, *A Contribution to the Critique of Political Economy* (Chicago, 1904), pp. 11–12.

20. As a reminder, we have only to read the sentence that follows the passage quoted from Marx in the text above. "It is not the consciousness of men that determines their [social] existence, but, on the contrary, their social existence determines their consciousness."

21. Friedrich Engels, "Letters . . . ," in Karl Marx, *Selected Works* (Moscow, 1936), vol. 1, p. 385.

22. Ibid., vol. 1, p. 386.

23. Ibid., vol. 1, p. 392.

24. Sorokin, *Dynamics*, vol. 2, pp. 474–75.

25. Engels, "Letters," 393.

26. As Sorokin informs us, he uses the term "science" as a shorthand expression for science and technology. In this passage he is evidently concerned with the low technology of everyday life in nonliterate societies rather than with science, strictly speaking.

27. Sorokin, *Dynamics*, vol. 4, p. 111.

28. Sorokin, *Society, Culture and Personality*, pp. 352–53n.

29. Maquet, *Sociology of Knowledge*, p. 202.

30. Merton, *Social Theory*, p. 559n. On a far less extensive scale, the use of quantitative indicators in the sociology of science will be found in R. K. Merton, *Science, Technology and Society in Seventeenth-Century England*, and in Nicholas Hans, *New Trends in Education in the Eighteenth Century*.

31. Sorokin, *Dynamics*, vol. 2, p. 125. All of chap. 3 is devoted to the presentation of such evidence.

32. Ibid., pp. 125–31.
33. Ibid., p. 439.
34. Sorokin, *Sociocultural Causality, Space, Time*, p. 95n.
35. Sorokin, *Dynamics*, vol. 2, p. 51.
36. For an excellent and intensive examination of the answer to this question implicit in Sorokin's work, see the paper by Riley and Moore in Philip J. Allen, ed., *Pitirim A. Sorokin in Review* (Durham, N.C.: Duke University Press, 1963).
37. Stark, *Sociology of Knowledge*, p. 280.
38. Sorokin, *Fads and Foibles*, pp. 160–61.
39. Sorokin, *Dynamics*, vol. 2, p. 20.
40. Einstein, *The World as I See It* (New York, 1934), p. 30.
41. Sorokin, *Dynamics*, vol. 2, p. 5.
42. Ibid., p. 9.
43. Ibid., p. 11n.
44. Ibid., pp. 11–12n.
45. Einstein, *The World as I See It*, pp. 36–37.
46. That is to say, Sorokin amply recognizes the fact but does not draw the possible implications of the fact for his theory. Thus: "The trend for the last four centuries has been for empiricism to rise steadily until, at the beginning of the twentieth century, it reached a unique, unprecedented [*n.b.*] level. . . . There was also a unique and unprecedented multiplication . . . of important discoveries and inventions in the sciences. Thus we truly live in the age of the truth of senses, of a magnitude, depth, and brillancy hardly witnessed in other cultures and periods" (*Dynamics*, vol. 2, p. 113).
47. Ibid., vol. 2, chap. 12.
48. Ibid., vol. 2, p. 467.
49. Robert K. Merton, "Singletons and Multiples in Scientific Discovery: A Chapter in the Sociology of Science," *Proceedings of the American Philosophical Society* 105, no. 5 (1961): 470–86.
50. Sorokin, *Dynamics*, vol. 2, pp. 119–20.
51. In less passionate prose than he employed in his lecture, Sorokin wrote in 1937: "Suppose someone should discover a simple but terrific explosive which could easily destroy a considerable part of our planet. Scientifically, it would be the greatest discovery, but socially the most dangerous for the very existence of mankind, because out of 1,800,000,000 human beings there certainly would be a few individuals who, being 'scientifically minded,' would like to test the explosive and as a result would destroy our planet. Such an explosion would be a great triumph of science. . . . This half-fantastic example shows that there must be limitations of science imposed by the reasons which are outside it, and these reasons usually come from the truth of faith and that of reason" (*Dynamics*, vol. 2, p. 20).
52. A. C. Crombie, *Augustine to Galileo: The History of Science A.D. 400–1650* (London, 1952), chaps. 1–4.
53. C. P. Snow, *The Two Cultures and the Scientific Revolution* (Cambridge: Cambridge University Rede Lectures, 1959), p. 1.
54. Ibid., p. 4.
55. Ibid., p. 15.
56. On the norm of humility in science, see Merton, "Priorities in Scientific Discovery: A Chapter in the Sociology of Science," *American Sociological Review* 22 (1957): 635–59, esp. pp. 646–47. The most famous expression of this norm by Newton can perhaps bear still another repetition: "I do not know what I may

appear to the world, but to myself I seem to have been only like a boy playing on the seashore, and diverting myself in now and then finding a smoother pebble or a prettier shell than ordinary, whilst the great ocean of truth lay all undiscovered before me."

57. Sorokin, *Dynamics*, vol. 2, p. 304.
58. Sorokin, *Society, Culture and Personality*, p. 30.

# Part II

# The Social Process of
# Scientific Discovery

# Introduction

Although it may seem obvious now, the process of scientific discovery was not conceived in the late nineteenth and early twentieth centuries as a *social* process, as much so as any other kind of social action. Science was thought to come just from science, through the action of remarkable individuals, often praised as "geniuses," who worked by themselves with unknown, strange, even mysterious modes of thought to arrive at often totally unexpected ("Eureka!") results. Now we know that the process of scientific discovery mixes all three of the following: individual training and talent, collegial interaction, and social-structural and cultural influences of many different kinds. This new knowledge was hard-won. It came not by doctrinaire fiat, nor by instant transformation of received ideas. New theories of action generally and of science particularly (see, e.g., Thomas Kuhn's *The Structure of Scientific Revolutions*) played their part, but so also did a considerable number of close-up studies of how scientists actually thought and behaved.

The papers here in part II are illustrative of the new kind of empirical study that demonstrated the social process of scientific discovery. The first paper (written jointly with Professor Renée Fox), "The Case of the Floppy-Eared Rabbits: An Instance of Serendipity Gained and Serendipity Lost," was written in the late 1950s and based on intensive interviews with two distinguished biomedical scientists, Drs. Aaron Kellner and Lewis Thomas, one of whom had experienced what we called "serendipity lost," the other "serendipity gained" in connection with the observation of an anomalous and unexpected phenomenon, floppiness in rabbits' ears. By comparing the experiences of the two scientists, we hoped to find the essential differences that led to relative success or failure. The paper brings out the way in which received scientific ideas (e.g., about cartilage in rabbits' ears), unexpected and anomalous phenomena, striving for scientific success, and the functions of teaching are all among the social factors that influence the process of scientific discovery and often result in what was coming to be known at the time as "serendipity," or "happy accidental discovery." Our paper shows that serendipity was not so much a matter of mere chance as of a set of regularized occurrences and routines for scientific discovery. Partly because of its title, this paper took the fancy of many scientists and thereby made it easier for them to see science as it really was.

The second paper (chap. 5), "Resistance by Scientists to Scientific Discovery," struck an even stronger chord among scientists and has become probably the most frequently cited of all my papers in the sociology of science. Based on historical examples, it challenged the current bedrock assumption that all scientists are open-minded individuals who readily see and acknowledge scientific novelty. When the paper was published in the widely read journal *Science*, I was deluged by hundreds of requests for reprints, by telephone calls, and even by personal visits. Altogether it was a new experience for me. But it turned out that many of those who communicated with me or who later cited the paper did not understand my main point. I said that "resistance" occurred when four things happened: (1) a discovery was announced; (2) the discovery was rejected by established scientists; (3) it could be shown that this rejection was due to the direct operation of cultural (e.g., received ideas) or social-structural (e.g., social-status differences) factors; and (4) the discovery was later accepted by established science. My best example among several, I thought, was the initial resistance to and later hailing of Mendel's discoveries.

Many of those who communicated with me or who later cited my paper defined "resistance" more broadly. They wanted to include all the critical scrutiny on good scientific grounds that is so essential a part of the social process of scientific discovery. I now realize that scientists are more emotionally involved with their own ideas than they are thought to be, that therefore even normal scientific criticism is often hard for them to bear. That is why they took my work for what it was not, a justification of their dislike of justifiable scientific criticism. Much scientific acrimony and conflict arise from the failure to distinguish between what I defined as resistance and what all scientists must often experience as critical scientific scrutiny.

Does "resistance" in my sense still occur? It is hard to know, since reports on supposedly resisted discoveries frequently do not meet my third criterion, namely, that the resistance is due to specifiable social-structural or cultural factors. Take the following report as an example:

> Truly innovative science is often—perhaps usually—accompanied by skepticism, dismissal, and/or disdain from the ranks of established expertise. That proposition receives surprisingly strong support from a study of the top-ranking papers from Britain's premier medical journals. Data from the Institute for Scientific Information's *Science Citation Index* (SCI) show that no less than four of the six papers most cited from *The Lancet* and the *British Medical Journal* during the years 1955–1988 record ideas that were initially rejected or disbelieved.[1]

Unfortunately, in its brief accounts of the reasons for initial rejection of the ideas contained in these now classic papers, there is no specification in this article of the social-structural or cultural factors that caused these rejections.

Without such specification, we do not know whether we are dealing with my concept of "resistance" or with justifiable, normal scientific criticism. It is important to understand the nature of both types to get the whole picture of scientific discovery.

The third and fourth papers in part II, "The Functions and Dysfunctions of 'Fashion' in Science: A Case for the Study of Social Change" (chap. 6) and "Trust in Science" (chap. 7), resulted from applications to the sociology of science of work I had done previously in general sociology.

In 1952 I published a paper (with Lyle Lobel) titled " 'Fashion' in Women's Clothes and the American Social System."[2] That empirical study convinced me that "fashion" was another one of those many much-used, everyday, sociologically vague terms that needed theoretical clarification and specification. I also saw that the term was used in reference to all sorts of social and cultural phenomena, and not least of all science itself.

I set out, therefore, to investigate what was called "fashion" in science and used the specific case of the recent rapid development of medical sociology, where empirical data were available, to make my investigation. As a result, I saw that what users of the term "fashion" in this case were referring to in a vague way was social and cultural change, *the sources, rates, and consequences of such change.* I said that it would be better to measure directly these sources, rates, and consequences than to continue with the vague notion of "fashion." On this understanding, the emergence and development of medical sociology as a scientific specialty was much more clearly comprehended than it was by simply being called a "fashion." Indeed, I decided that the term "fashion" needed to be abandoned for any good sociological analysis. Unfortunately, my conviction has not been accepted and the loose usage of the term persists nearly everywhere.

The last paper in part II, on trust in science, was written for a symposium in honor of the late, distinguished sociologist of science, and educator, Professor Joseph Ben-David. Again, it was an application of some more general sociological theory and data developed in a book on trust.[3] That theory had argued that all social action is based on expectations of what I called technical competence and/or fiduciary responsibility. "Trust" is often used to refer to one or the other, or both. Complete trust is where both are present. Nowhere is trust more necessary than in science, both within science itself and in the relations between science and the rest of society. Much of the present controversies over such matters as the alleged "commercialization" of, "fraud" in, and "mortal danger" from science result from the breakdown of internal and external trust in science. As science continues to develop and grow more powerful in its social and cultural consequences, we shall have to attend ever more closely to the problem of trust in science. In part III, "The Dilemma of

Science and Therapy," we shall consider further the problem of trust, namely, trust in biomedical scientists who use human subjects in their research.

## Notes

1. *The Scientist*, April 17, 1989, p. 12.
2. *Social Forces* 31 (1952): 124–131.
3. Bernard Barber, *The Logic and Limits of Trust* (New Brunswick, N.J.: Rutgers University Press, 1983).

# 4

# The Case of the Floppy-Eared Rabbits:
# An Instance of Serendipity Gained
# and Serendipity Lost

*Bernard Barber and Renée C. Fox*

As with so many other basic social processes, the actual process of scientific research and discovery is not well understood.[1] There has been little systematic observation of the research and discovery process as it actually occurs, and even less controlled research. Moreover, the form in which discoveries are reported by scientists to their colleagues in professional journals tends to conceal important aspects of this process. Because of certain norms that are strongly insititutionalized in their professional community, scientists are expected to focus their reports on the logical structure of the methods used and the ideas discovered in research in relation to the established conceptual framework of the relevant scientific specialty. The primary function of such reports is conceived to be that of indicating how the new observations and ideas being advanced may require a change—by further generalization or systematization—in the conceptual structure of a given scientific field. All else that has occurred in the actual research process is considered "incidental." Thus scientists are praised for presenting their research in a way that is elegantly bare of anything that does not serve this primary function and are deterred from reporting "irrelevant" social and psychological aspects of the research process, however interesting these matters may be in other contexts. As a result of such norms and practices, the reporting of scientific research may be characterized by what has been called "retrospective falsification." By selecting only those components of the actual research process that serve their primary purpose, scientific papers leave out a great deal, of course, as many scientists have indicated in their memoirs and in their informal talks with one another. Selection, then, unwittingly distorts and, in that special sense, falsifies what has happened in research as it actually goes on in the laboratory and its environs.

Public reports to the community of scientists thus have their own function. Their dysfunctionality for the sociology of scientific discovery, which is concerned with not one but all the components of the research process as a social process, is of no immediate concern to the practicing research scientist. And yet what is lost in "retrospective falsification" may be of no small importance to him, if only indirectly. For it is not unlikely that here, as everywhere else in the world of nature, knowledge is power, in this case power to increase the fruitfulness of scientific research by enlarging our systematic knowledge of it. The sociology of scientific discovery would seem to be an especially desirable area for further theoretical and empirical development.

One component of the actual process of scientific discovery that is left out or concealed in research reports following the practice of "retrospective falsification" is the element of unforeseen development, of happy or lucky chance, of what Robert K. Merton has called "the serendipity pattern."[2] By its very nature, scientific research is a voyage into the unknown by routes that are in some measure unpredictable and unplannable. Chance or luck is therefore as inevitable in scientific research as are logic and what Pasteur called "the prepared mind." Yet little is known systematically about this inevitable serendipity component.

For this reason it seemed to us desirable to take the opportunity recently provided by the reporting of an instance of *serendipity gained* by Dr. Lewis Thomas, now professor and chairman of the Department of Medicine in the College of Medicine of New York University and formerly professor and chairman of the Department of Pathology.[3] Then, shortly after hearing about Dr. Thomas's discovery, we learned from medical research and teaching colleagues of an instance of *serendipity lost* on the very same kind of chance occurrence: unexpected floppiness in rabbits' ears after they had been injected intravenously with the proteolytic enzyme papain. This instance of serendipity lost had occurred in the course of research by Dr. Aaron Kellner, associate professor in the Department of Pathology of Cornell University Medical College and director of its central laboratories. This opportunity for *comparative* study seemed even more promising for our further understanding of the serendipity pattern. Here were two comparable medical scientists, we reasoned, both carrying out investigations in the field of experimental pathology, affiliated with distinguished medical schools, and of approximately the same level of demonstrated research ability (so far as it was in our lay capacity to judge). In the course of their research both men had had occasion to inject rabbits intravenously with papain, and both had observed the phenomenon of ear collapse following the injection.

In spite of these similarities in their professional backgrounds and although they had both accidentally encountered the same phenomenon, one of these scientists had gone on to make a discovery based on this chance occurrence, whereas the other had not. It seemed to us that a detailed comparison of Dr. Thomas's and Dr. Kellner's experiences with the floppy-eared rabbits offered a quasi-experimental opportunity to identify some of the factors that contribute to a positive experience with serendipity in research and some of the factors conducive to a negative experience with it.

We asked for and were generously granted intensive interviews with Dr. Thomas and Dr. Kellner.[4] Each reported to us that they had experienced both "positive serendipity" and "negative serendipity" in their research. That is, each had made a number of serendipitous discoveries based on chance occurrences in their planned experiments, and on other occasions each had missed the significance of like occurrences that other researchers had later transformed into discoveries. Apparently, both positive and negative serendipity are common experiences for scientific researchers. Indeed, we shall see that one of the chief reasons why Dr. Kellner experienced serendipity lost with respect to the discovery that Dr. Thomas made was that he was experiencing serendipity gained with respect to some other aspects of the very same experimental situation. Conversely, Dr. Thomas had reached a stalemate on some of his other research, and this gave him added incentive to pursue intensively the phenomenon of ear collapse. Partly as a consequence of these experiences, in what were similar experimental situations, the two researchers each saw something and missed something else.

On the basis of our focused interviews with these two scientists, we can describe some of the recurring elements in their experiences with serendipity.[5] We think that these patterns may also be relevant to instances of serendipity experienced by other investigators.

### Serendipity Gained

*Dr. Thomas*—Observing the established norms for reporting scientific research, in his article in the *Journal of Experimental Medicine*, Dr. Thomas did not mention his experience with serendipity. In the manner typical of such reports he began his article with the statement, "For reasons not relevant to the present discussion rabbits were injected intravenously with a solution of crude papain." (By contrast, though not called by this term, serendipity was featured in the accounts of this research that appeared in the *New York Times* and the *New York Herald Tribune*. "An accidental sidelight of one research project had the startling effect of wilting the ears of the rabbit," said the *Times* article. "This bizarre phenomenon, accidentally discovered . . ." was the way the *Herald Tribune* described the same phenomenon. The prominence accorded the "accidental" nature of the discovery in the press is related to the fact that these articles were written by journalists for a lay audience. The kind of interest in scientific research that is characteristic of science reporters and the audience for whom they write and their conceptions of the form in which information about research ought to be communicated differ from those of professional scientists.)[6]

Although Dr. Thomas did not mention serendipity in his article for the *Journal of Experimental Medicine*, in his interview he reported both his general acquaintance with the serendipity pattern ("Serendipity is a familiar term. . . . I first heard about it in Dr. Cannon's class. . . .") and his awareness of the chance occurrence of floppy-eared rabbits in his own research. Dr. Thomas had first noticed the reversible collapse of rabbit ears after intravenous papain about seven years earlier when he was working on the effects of proteolytic enzymes as a class:

> I was trying to explore the notion that the cardiac and blood vessel lesions in certain hypersensitivity states may be due to release of proteolytic enzymes. It's an attractive idea on which there's little evidence. And it's been picked up at some time or another by almost everyone working on hypersensitivity. For this investigation I used trypsin, because it was the most available enzyme around the laboratory, and I got nothing. We also happened to have papain; I don't know where it had come from; but because it was there, I tried it. I also tried a third enzyme, ficin. It comes from figs, and it's commonly used. It has catholic tastes and so it's quite useful in the laboratory. So I had these three enzymes. The other two didn't produce lesions. Nor did papain. But what the papain did was always produce these bizarre cosmetic changes. . . . It was one of the most uniform reactions I'd ever seen in biology. It always happened. And it looked as if something important must have happened to cause this reaction.

Some of the elements of serendipitous discovery are clearly illustrated in this account by Dr. Thomas. The scientific researcher, while in pursuit of some other specific goals, accidentally ("we also happened to have papain . . .") produces an unusual, recurrent, and sometimes striking ("bizarre") effect. Only the element of creative imagination, which is necessary to complete an instance of serendipity by supplying an explanation of the unusual effect, is not yet present. Indeed, the explanation was to elude Dr. Thomas, as it eluded Dr. Kellner, and probably others as well, for several years. This was not for lack of trying by Dr. Thomas. He immediately did seek an explanation:

> I chased it like crazy. But I didn't do the right thing. . . . I did the expected things. I had sections cut, and I had them stained by all the techniques available at the time.

And I studied what I believed to be the constituents of a rabbit's ear. I looked at all the sections, but I couldn't see anything the matter. The connective tissue was intact. There was no change in the amount of elastic tissue. There was no inflammation, no tissue damage. I expected to find a great deal, because I thought we had destroyed something.

Dr. Thomas also studied the cartilage of the rabbit's ear, and judged it to be "normal" (". . . The cells were healthy-looking and there were nice nuclei. I decided there was no damage to the cartilage. And that was that . . ."). However, he admitted that at the time his consideration of the cartilage was routine and relatively casual, because he did not seriously entertain the idea that the phenomenon of ear collapse might be associated with changes in this tissue:

I hadn't thought of cartilage. You're not likely to, because it's not considered interesting. . . . I know my own idea has always been that cartilage is a quiet, inactive tissue.

Dr. Thomas's preconceptions about the methods appropriate for studying the ear-collapsing effect of papain, his expectation that it would probably be associated with damage in the connective or elastic tissues, and the conviction he shared with colleagues that cartilage is "inert and relatively uninteresting"—these guided his initial inquiries into this phenomenon. But the same preconceptions, expectations, and convictions also blinded him to the physical and chemical changes in the ear-cartilage matrix, which, a number of years later, were to seem "obvious" to him as the alterations underlying the collapsing ears. Here again, another general aspect of the research process comes into the clear. Because the methods and assumptions on which a systematic investigation is built selectively focus the researcher's attention, to a certain extent they sometimes constrict his imagination and bias his observations.

Although he was "very chagrined" about his failure, Dr. Thomas finally had to turn away from his floppy-eared rabbits because he was "terribly busy working on another problem at the time," with which he was "making progress." Also, Dr. Thomas reported, "I had already used all the rabbits I could afford. So I was able to persuade myself to abandon this other research." The gratifications of research success elsewhere and the lack of adequate resources to continue with his rabbit experiments combined to make Dr. Thomas accept failure, at least temporarily. As is usually the case in the reporting of scientific research, these experiments and their negative outcome were not written up for professional journals. (There is too much failure of this sort in research to permit of its publication, except occasionally, even though it might be instructive for some other scientists in carrying out their research. Since there is no way of determining what might be instructive failures and since space in professional journals is at a premium, generally only accounts of successful experiments are submitted to such journals and published by them.)

Despite his decision to turn his attention to other, more productive research, Dr. Thomas did not completely forget the floppy-eared rabbits. His interest was kept alive by a number of things. As he explained, the collapse of the rabbit ears and their subsequent reversal "was one of the most uniform reactions I'd ever seen in biology." The "unfailing regularity" with which it occurred is not often observed in scientific research. Thus the apparent invariance of this phenomenon never ceased to intrigue Dr. Thomas, who continued to feel that an important and powerful biological happening might be responsible. The effect of papain on rabbit ears had two additional qualities that helped to sustain Dr. Thomas's interest in it. The spectacle of rabbits with "ears collapsed limply at either side of the head, rather like the ears of spaniels,"[7] was both dramatic and entertaining.

In the intervening years Dr. Thomas described this phenomenon to a number of colleagues in pathology, biochemistry, and clinical investigation, who were equally intrigued and of the opinion that a significant amount of demonstrable tissue damage must be associated with such a striking and uniform reaction. Dr. Thomas also reported that twice he "put the experiment on" for some of his more skeptical colleagues. ("They didn't believe me when I told them what happened. They didn't really believe that you can get that much change and not a trace of anything having happened when you look in the microscope.") As so often happens in science, an unsolved puzzle was kept in mind for eventual solution through informal exchanges between scientists, rather than through the formal medium of published communications.

A few years ago Dr. Thomas once again accidentally came upon the floppy-eared rabbits in the course of another investigation:

> I was looking for a way. . . to reduce the level of fibrinogen in the blood of rabbits. I had been studying a form of fibrinoid which occurs inside blood vessels in the generalized Schwartzman reaction and which seems to be derived from fibrinogen. My working hypothesis was that if I depleted the fibrinogen and, as a result, fibrinoid did not occur, this would help. It had been reported that if you inject proteolytic enzyme, this will deplete fibrinogen. So I tried to inhibit the Schwartzman reaction by injecting papain intravenously into the rabbits. It didn't work with respect to fibrinogen. . . . But the same damned thing happened again to the rabbits' ears!

This time, however, Dr. Thomas was to solve the puzzle of the collapsed rabbits' ears and realize a complete instance of serendipitous discovery. He describes what subsequently happened:

> I was teaching second-year medical students in pathology. We have these small seminars with them: two-hour sessions in the morning, twice a week, with six to eight students. These are seminars devoted to experimental pathology and the theo-

retical aspects of the mechanism of disease. The students have a chance to see what we, the faculty, are up to in the laboratory. I happened to have a session with the students at the same time that this thing with the rabbits' ears happened again. I thought it would be an entertaining thing to show them . . . a spectacular thing. The students were very interested in it. I explained to them that we couldn't really explain what the hell was going on here. I did this experiment on purpose for them, to see what they would think. . . . Besides which, I was in irons on my other experiments. There was not much doing on those. I was not being brilliant on these other problems. . . . Well, this time I did what I didn't do before. I simultaneously cut sections of the ears of rabbits after I'd given them papain *and* sections of normal ears. This is the part of the story I'm most ashamed of. It still makes me writhe to think of it. There was no damage to the tissue in the sense of a lesion. But what had taken place was a quantitative change in the matrix of the cartilage. The only way you could make sense of this change was simultaneously to compare sections taken from the ears of rabbits which had been injected with papain with comparable sections from the ears of rabbits of the same age and size which had not received papain. . . . Before this I had always been so struck by the enormity of the change that when I didn't see something obvious, I concluded there was nothing. . . . Also, I didn't have a lot of rabbits to work with before.

Judging from Dr. Thomas's account, it appears that a number of factors contributed to his reported experimental success. First, his teaching duties played a creative role in this regard. They impelled him to run the experiment with papain again and kept his attention focused on its implications for basic science rather than on its potentialities for practical application. Dr. Thomas said that he used the experiment to "convey to students what experimental pathology is like." Second, because he had reached an impasse in some of his other research, Dr. Thomas had more time and further inclination to study the ear-collapsing effect of papain than he had had a few years earlier, when the progress he was making on other research helped to "persuade" him to "abandon" the problem of the floppy-eared rabbits. Third, Dr. Thomas had more laboratory resources at his command than previously, notably a larger supply of rabbits. (In this regard it is interesting to note that, according to Dr. Thomas's article in the *Journal of Experimental Medicine*, 250 rabbits, all told, were used in the experiments reported.) Finally, the fact that he now had more laboratory animals with which to work and that he wanted to present the phenomenon of reversible ear collapse to students in a way that would make it an effective teaching exercise led Dr. Thomas to modify his method for examining rabbit tissues. In his earlier experiments, Dr. Thomas had compared histological sections made of the ears of rabbits who had received an injection of papain with his own mental image of normal rabbit-ear tissue. This time, however, he actually made sections from the ear tissue of rabbits that did *not* receive papain, as well as from those that did, and simultaneously examined the two. As he reported, this comparison enabled him to see for the first time that "drastic" quantitative changes had occurred in the cartilaginous tissue obtained from the ears of the rabbits injected with papain. In the words of the *Journal* article:

> The ear cartilage showed loss of a major portion of the intercellular matrix, and complete absence of basophilia from the small amount of remaining matrix. The cartilage cells appeared somewhat larger, and rounder than normal, and lay in close contact with each other. . . .

Immediately thereafter, Dr. Thomas and his associates found that these changes occur not only in ear cartilage but in all other cartilaginous tissues as well.

How significant or useful Dr. Thomas's serendipitous discovery will be cannot yet be specified. The serendipity pattern characterizes small discoveries as well as great. Dr. Thomas and his associates are currently investigating some of the questions raised by the phenomenon of papain-collapsed ears and the alterations in cartilage now known to underlie it. In addition, Dr. Thomas reported that some of his "biochemist and clinical friends" have become interested enough in certain of his findings to "go to work with papain, too." Two of the major problems under study in Dr. Thomas's laboratory are biochemical: the one concerning the nature of the change in cartilage; the other, the nature of the factor in papain that causes collapse of rabbits' ears and lysis of cartilage matrix in all tissues. Attempts are also being made to identify the antibody that causes rabbits to become immune to the factor responsible for ear collapse after two weeks of injection. The way in which cortisone prolongs the reaction to papain and the possible effect that papain may have on the joints as well as the cartilage are also being considered. Though at the time he was interviewed Dr. Thomas could not predict whether his findings (to date) would prove "important" or not, there was some evidence to suggest that certain basic discoveries about the constituents and properties of cartilaginous tissue might be forthcoming and that the experiments thus far conducted might have "practical usefulness" for studies of the postulated role of cortisone in the metabolism of sulfated mucopolysaccharides and of the relationship between cartilage and the electrolyte imbalance associated with congestive heart failure.

In the research on reversible ear collapse that Dr. Thomas has conducted since his initial serendipitous discovery, the planned and the unplanned, the foreseen and the accidental, the logical and the lucky have continued to interact. For example, Dr. Thomas's discovery that cortisone prevents or greatly delays the "return of papain-collapsed ears to their normal shape and rigidity" came about as a result of a carefully planned experiment that he undertook to test the effect of cortisone on the reaction to papain. On the other hand, his discovery that "repeated injections of papain, over a period of two or three weeks, brings about immunity to the phenomenon of ear collapse" was an unanticipated consequence of the fact that he used the same rabbit to demonstrate the floppy ears to several different groups of medical students:

> I was so completely sold on the uniformity of this thing that I used the same rabbit [for each seminar]. . . . The third time it didn't work. I was appalled by it. The students were there, and the rabbit's ears were still in place. . . . At first I thought that perhaps the technician had given him the wrong stuff. But then when I checked on that and gave the same stuff to the other rabbits and it *did* work I realized that the rabbit had become immune. This is a potentially hot finding. . . .

## Serendipity Lost

*Dr. Kellner.*—In our interview with Dr. Thomas we told him that we had heard about another medical scientist who had noticed the reversible collapse of rabbits' ears when he had injected them intravenously with papain. Dr. Thomas was not at all surprised. "That must be Kellner," he said. "He must have seen it. He was doomed to see it." Dr. Thomas was acquainted with the reports that Dr. Kellner and his associates had published on "Selective Necrosis of Cardiac and Skeletal Muscle Induced Experimentally by Means of Proteolytic Enzyme Solutions Given Intravenously" and on "Blood Coagulation Defect Induced in Rabbits by Papain Solutions Injected Intravenously."[8] He took it for granted that, in the course of these reported experiments, which had entailed papain solution given intravenously to rabbits, a competent scientist like Dr. Kellner had also seen the resulting collapse of rabbits' ears, with its "unfailing regularity" and its "flamboyant" character. And, indeed, our interview with Dr. Kellner revealed that he had observed the floppiness, apparently at about the same time as Dr. Thomas:

> We called them the floppy-eared rabbits. . . . Five or six years ago we published our first article on the work we were doing with papain; that was in 1951 and our definitive article was published in 1954. . . . We gave papain to the animals and we had done it thirty or forty times before we noticed these changes in the rabbits' ears.

Thus Dr. Kellner's observation of what he and his colleagues dubbed "the floppy-eared rabbits" represents, when taken together with Dr. Thomas's experience, an instance of independent multiple observation, which often occurs in science and frequently leads to independent multiple invention and discovery.

Once he had noticed the phenomenon of ear collapse, Dr. Kellner did what Dr. Thomas and any research scientist would have done in the presence of such an unexpected and striking regularity: he looked for an answer to the puzzle it represented. "I was a little curious about it at the time, and followed it up to the extent of making sections of the rabbits' ears." However, for one of those trivial reasons that sometimes affect the course of research—the obviously amusing quality of floppiness in rabbits' ears—Dr. Kellner did not take the phenomenon as seriously as he took other aspects of the experimental situation involving the injection of papain.

In effect, Dr. Kellner and his associates closed out their interest in the phenomenon of the reversible collapse of rabbits' ears following intravenous injection of papain by using it as an assay test for the potency and amount of papain to be injected. "Every laboratory technician we've had since 1951," he told us in the interview, "has known about these floppy ears because we've used them to assay papain, to tell us if it's potent and how potent." If the injected rabbit died from the dose of papain he received, the researchers knew that the papain injection was too potent; if there was no change in the rabbit's ears, the papain was not potent enough, but "if the rabbit lived and his ears drooped, it was just right." Although "we knew all about it, and used it that way. . . as a rule of thumb," Dr. Kellner commented, "I didn't write it up." Nor did he ever have "any intention of publishing it as a method of assaying papain." He knew that an applied technological discovery of this sort would not be suitable for publication in the basic science-oriented professional journals to which he and his colleagues submit reports of experimental work.

However, two factors apparently were much more important in leading Dr. Kellner away from investigating this phenomenon. First, like Dr. Thomas, Dr. Kellner thought of cartilage as relatively inert tissue. Second, because of his preestablished special research interests, Dr. Kellner's attention was predominantly trained on muscle tissue:

> Since I was primarily interested in research questions having to do with the muscles of the heart, I was thinking in terms of muscle. That blinded me, so that changes in the cartilage didn't occur to me as a possibility. I was looking for muscles in the sections, and I never dreamed it was cartilage.

Like Dr. Thomas at the beginning of his research and like all scientists at some stages in their research, Dr. Kellner was "misled" by his preconceptions.

However, as we already know, in keeping with his special research interests, Dr. Kellner noticed and intensively followed up two other serendipitous results that occur when papain is injected intravenously into rabbits: focal necrosis of cardiac and skeletal muscle and a blood coagulation defect, which in certain respects resembles that of hemophilia.[9]

It was the selective necrosis of cardiac and skeletal muscle that Dr. Kellner studied with the greatest degree of seriousness and interest. Dr. Kellner told us that he is "particularly interested in cardio-vascular disease," and so the lesions in the myocardium was the chance observation that he particularly "chose to follow . . . the one closest to me." Not only did Dr. Kellner himself have a special interest in the necrosis of cardiac muscle, but also his "laboratory and the people associated with me," he said, provided "the physical and intellectual tools to cope with this phenomenon." Dr. Kellner and his colleagues also did a certain amount of "work tracking down the cause of the blood coagulation defect"; but, because this line of inquiry "led [them] far afield" from investigative work in which they were especially interested and competent, they eventually "let that go" as they had let go the phenomenon of floppiness in rabbits' ears. Dr. Kellner indicated in his interview that the potential usefulness of his work with the selective necrosis of cardiac and skeletal muscle cannot yet be precisely ascertained. However, in his article in the *Journal of Experimental Medicine* he suggested that this serendipitous finding "has interesting implications for the pathogenesis of the morphological changes in rheumatic fever, periarteritis nodosa, and other hypersensitivity states."

Thus Dr. Kellner did not have the experience of serendipity gained with respect to the significance of floppiness in rabbits' ears after intravenous injection of papain for a variety of reasons, some trivial apparently, others important. The most important reasons, it seems, were his research preconceptions and the occurrence of other serendipitous phenomena in the same experimental situation.

In summary, although the ultimate outcome of their respective laboratory encounters with floppiness in rabbits' ears was quite different, there are some interesting similarities between the serendipity-gained experience of Dr. Thomas and the serendipity-lost experience of Dr. Kellner. Initially, the attention of both men was caught by the striking uniformity with which the collapse of rabbit ears occurred after intravenous papain and by the "bizarre," entertaining qualities of this cosmetic effect. In their subsequent investigations of this phenomenon, both were to some extent misled by certain of their interests and preconceptions. Lack of progress in accounting for ear collapse, combined with success in other research in which they were engaged at the time, eventually led both Dr. Thomas and Dr. Kellner to discontinue their work with the floppy-eared rabbits.

However, there were also some significant differences in the two experiences. Dr. Thomas seems to have been more impressed with the regularity of this particular phenomenon than Dr. Kellner and somewhat less amused by it. Unlike Dr. Kellner, Dr. Thomas never lost interest in the floppy-eared rabbits. When he came upon this reaction again at a time when he was "blocked" on other research, he began actively to reconsider the problem of what might have caused it. Eventual success was more likely to result from this continuing concern on Dr. Thomas's part. And, Dr. Kellner, of course, was drawn off in other research directions by seeing other serendipitous phenomena in the same situation and by his success in following up those other leads.

These differences between Dr. Thomas and Dr. Kellner seem to account at least in part for the serendipity-gained outcome of the case of the floppy-eared rabbits for the one, and the serendipity-lost outcome for the other.

Experiences with both serendipity gained and serendipity lost are probably frequent occurrences for many scientific researchers. For, as Dr. Kellner pointed out in our interview with him, scientific investigations often entail "doing something that no one has done before, [so] you don't always know how to do it or exactly what to do":

> Should you boil or freeze, filter or centrifuge? These are the kinds of crossroads you come to all the time. . . . It's always possible to do four, five, or six things, and you have to choose between them. . . . How do you decide?

In this comparative study of one instance of serendipity gained and serendipity lost, we have tried to make inferences about some of the factors that led one investigator down the path to a successful and potentially important discovery and another to follow a somewhat different, though eventually perhaps a no less fruitful, trail of research. A large enough series of such case studies could suggest how often and in what ways these factors (and others that might prove relevant) influence the paths that open up to investigators in the course of their research, the choices they make between them, and the experimental findings that result from such choices. Case studies of this kind might also contribute a good deal to the detailed, systematic study of "the ways in which scientists actually. . . think, feel and act," which Robert K. Merton says could perhaps teach us more "in a comparatively few years, about the psychology and sociology of science than in all the years that have gone before."[10]

## Notes

1. For an account of what is known, see Bernard Barber, *Science and the Social Order* (Glencoe, Ill.: Free Press, 1952), chap. 9, "The Social Process of Invention and Discovery," pp. 191–206.
2. For discussions of serendipity, see Walter B. Cannon, *The Way of an Investigator* (New York: W. W. Norton & Co., 1945), chap. 6, "Gains from Serendipity," pp. 68–78; and Robert K. Merton, *Social Theory and Social Structure* (rev. ed.; Glencoe, Ill.: Free Press, 1957), pp. 103–8. Our colleagues, Robert K. Merton and Elinor G. Barber, are now engaged in an investigation and clarification of the variety of meanings of "chance" that are lumped under the notion of serendipity by different users of that term.
3. Lewis Thomas, "Reversible Collapse of Rabbit Ears after Intravenous Papain, and Prevention of Recovery by Cortisone," *Journal of Experimental Medicine* 104 (1956): 245–52. This case first came to our attention through a report in the

*New York Times*. The pictures printed in Dr. Thomas's original article and in the *Times* will indicate why we have called this "the case of the floppy-eared rabbits."

4. These interviews lasted about two hours each. They are another instance of the "tandem interviewing" described by Harry V. Kincaid and Margaret Bright, "Interviewing the Business Elite," *American Journal of Sociology* 63 (1957): 304–11.

5. In this paper we shall concentrate on the instances of serendipity gained by Dr. Thomas and lost by Dr. Kellner and give somewhat less attention to elements of negative serendipity in Dr. Thomas's experiments and elements of positive serendipity in those of Dr. Kellner.

6. Further discussion of this point lies beyond the scope of this paper. But in a time like ours, in which science has become "front-page news," some of the characteristics and special problems of science reporting merit serious study. A published work on this topic that has come to our attention is entitled *When Doctors Meet Reporters* (New York: New York University Press, 1957). This is a discussion by science writers and physicians of the controversy between the press and the medical profession, compiled from the record of a series of conferences sponsored by the Josiah Macy, Jr., Foundation.

7. Thomas, "Reversible Collapse of Rabbit Ears," p. 245.

8. See Aaron Kellner and Theodore Robertson, "Selective Necrosis of Cardiac and Skeletal Muscle Induced Experimentally by Means of Proteolytic Enzyme Solutions Given Intravenously," *Journal of Experimental Medicine* 99 (1954): 387–404; and Aaron Kellner, Theodore Robertson, and Howard O. Mott, "Blood Coagulation Defect Induced in Rabbits by Papain Solutions Injected Intravenously," abstract in *Federation Proceedings* 10 (1951), no. 1.

9. See Kellner and Robertson, "Selective Necrosis . . . ," and Kellner, Robertson, and Mott, "Blood Coagulation Defect. . . ."

10. See Merton, Foreword, in Barber, *Science and the Social Order*, p. xxii.

# 5

# Resistance by Scientists to Scientific Discovery

In the study of the history and sociology of science, there has been a relative lack of attention to one of the interesting aspects of the social process of discovery—the resistance on the part of scientists themselves to scientific discovery. General and specialized histories of science and biographies and autobiographies of scientists, as well as intensive discussions of the processes by which discoveries are made and accepted, all tend to make, at the most, passing reference to this subject. In two systematic analyses of the social process of scientific discovery and invention, for example—analyses which tried to be as inclusive of empirical fact and theoretical problems as possible—there is only passing reference to such resistance, in the one instance, and none at all in the second.[1] This neglect is all the more notable in view of the close scrutiny that scholars have given the subject of resistance to scientific discovery by social groups other than scientists. There has been a great deal of attention paid to resistance on the part of economic, technological, religious, and ideological elements and groups outside science itself.[1-3] Indeed, the tendency of such elements to resist seems sometimes to be emphasized disproportionately as against the support that they also give to science. In the matter of religion, for example, are we not all a little too much aware that religion has resisted scientific discovery, not enough aware of the large support it has given to Western science?[4-5]

The mere assertion that scientists themselves sometimes resist scientific discovery clashes, of course, with the stereotype of the scientist as "the open-minded man." The norm of open-mindedness is one of the strongest of the scientist's values. As Philipp Frank put it, "Every influence of moral, religious, or political considerations upon the acceptance of a theory is regarded as 'illegitimate' by the so-called 'community of scientists.'" And Robert Oppenheimer emphasizes the "importance" of "the open mind," in a book by that title, as a value not only for science but for society as a whole.[6] But values alone, and especially one value by itself, cannot be a sufficient basis for explaining human behavior. However strong a value is, however large its actual influence on behavior, it usually exerts this influence only in conjunction with a number of other cultural and social elements, which sometimes reinforce it, sometimes give it limits.

This article is an investigation of the elements within science that limit the norm and practice of "open-mindedness." My purpose is to draw a more accurate picture of the actual process of scientific discovery, to see resistance by scientists themselves as a constant phenomenon with specifiable cultural and social sources. This purpose, moreover, implies a practical consequence. For if we learn more about resistance to scientific discovery, we shall know more also about the sources of acceptance, just as we know more about health when we successfully study disease. By knowing more about both resistance and acceptance in scientific discovery, we may be able to reduce the former by a little bit and thereby increase the latter in the same measure.

### Helmoltz, Planck, and Lister

Although the resistance by scientists themselves to scientific discovery has been neglected in systematic analysis, it would be surprising indeed if it had never been noted at all. If nowhere else, we should find it in the writings of those scientists who have suffered from resistance on the part of other scientists. Helmholtz, for example, made aware of such resistance by his own experience, commiserated with Faraday on "the fact that the greatest benefactors of mankind usually do not obtain a full reward during their life-time, and that new ideas need the more time for gaining general assent the more really original they are".[7-9] Max Planck is another who noticed resistance in general because he had experienced it himself, in regard to some new ideas on the second law of thermodynamics, which he worked out in his doctoral dissertation submitted to the University of Munich in 1879. Ironically, one of those who resisted the ideas proposed in Planck's paper, according to his account, was Helmholtz: "None of my professors at the University had any understanding for its contents," says Planck:

> I found no interest, let alone approval, even among the very physicists who were closely connected with the topic. Helmholtz probably did not even read my paper at all. Kirchhoff expressly disapproved, . . . I did not succeed in reaching Clausius. He did not answer my letter, and I did not find him at home when I tried to see him in person at Bonn. I carried on a correspondence with Carl Neumann, of Leipzig, but it remained totally fruitless.[10], [p.18]

And Lister, in a graduation address to medical students, warned them all against blindness to new ideas in science, blindness such as he had encountered in advancing his theory of antisepsis.

## Scientists Are Also Human

Too often, unfortunately, where resistance by scientists has been noted, it has been merely noted, merely alleged, without detailed substantiation and without attempt at explanation. Sometimes, when explanations are offered, they are notably vague and all-inclusive, thus proving too little by trying to prove too much. One such explanation is contained in the frequently repeated phrase, "After all, scientists are also human beings," words implying that scientists are more human when they err than when they are right."[11] Other vague explanations can be found in phrases such as "*Zeitgeist*," "human nature," "lack of progressive spirit," "fear of novelty," and "climate of opinion."

As one of these phrases, "fear of novelty," may indicate, there has also been a tendency, where some explanation of the sources of resistance is offered, to express a psychologistic bias—that is, to attribute resistance exclusively to inherent and ineradicable traits or instincts of the human personality. Thus Wilfred Trotter, in discussing the response to scientific discovery, asserts that "the mind delights in a static environment," that "change from without . . . seems in its very essence to be repulsive and an object of fear," and that "a little self-examination tells us pretty easily how deeply rooted in the mind is the fear of the new."[12] And Beveridge, in *The Art of Scientific Investigation*, says, "there is in all of us a psychological tendency to resist new ideas."[13] A full understanding of resistance will, of course, have to include the psychological dimension—the factor of individual personality. But it must also include the cultural and social dimensions—those shared and patterned idea-systems and those patterns of social interaction that also contribute to resistance. It is these cultural and social elements that I shall discuss here, but with full awareness that psychological elements are contributory causes of resistance.

Because resistance by scientists has been largely neglected as a subject for systematic investigation, we find that there is sometimes a tendency, when such resistance is noted, to exaggerate the extent to which it occurs. Thus, Murray says that the discoverer must *always* expect to meet with opposition from his fellow scientists. And Trotter goes overboard in the same way: "the reception of new ideas tends always to be grudging or hostile. . . . Apart from the happy few whose work has already great prestige or lies in fields that are being actively expanded at the moment, discoverers of new truths always find their ideas resisted"[12],p.26. Such exaggerations can be eliminated by more systematic and objective study.

Finally, in the absence of such systematic and objective study, many of those who have noted resistance have been excessively embittered and moralistic. Oliver Heaviside is reported to have exclaimed bitterly, when his important contributions to mathematical physics were ignored for twenty-five years, "Even men who are not Cambridge mathematicians deserve justice."[14] And Planck's reaction to the resistance he experienced was similar. "This experience," he said, "gave me also an opportunity to learn a new fact—a remarkable one, in my opinion: A new scientific truth does not triumph by convincing its opponents and making them see the light, but rather because its opponents eventually die, and a new generation grows up that is familiar with it."[10] Such bitterness is not tempered by objective understanding of resistance as a constant phenomenon in science, a pattern in which all scientists may sometimes and perhaps often participate, now on the side of the resisters, now on that of the resisted. Instead, such bitterness takes the moralistic view that resistance is due to "human vanities," to "little minds and ignoble minds." Such views impede the objective analysis that is required.

In his discussion of the idols—idols of the tribe, of the cave, of the marketplace, and of the theater—Francis Bacon long ago suggested that a variety of preconceived ideas, general and particular, affect the thinking of all men, especially in the face of innovation. Similarly, more recent sociological theory has shown that while the variety of idea-systems that make up a given culture are functionally necessary, on the whole, for man to carry on his life in society and in the natural environment, these several idea-systems may also have their dysfunctional or negative effects. Just because the established culture defines the situation for man, usually helpfully, it also, sometimes harmfully, blinds him to other ways of conceiving that situation. Cultural blinders are one of the constant sources of resistance to innovations of all kinds. And scientists, for all the methods they have invented to strip away their distorting idols, or cultural blinders, and for all the training they receive in evading the negative effects of such blinders, are still as other men, though surely in considerably lesser measure because of these methods and this special training. Scientists suffer, along with the rest of us, from the ironies that evil sometimes comes from good, that one noble vision may exclude another, and that good scientific ideas occasionally obstruct the introduction of better ones.

## Substantive Concepts

Several different kinds of cultural resistance to discovery may be distinguished. We may turn first to the way in which the substantive concepts and theories held by scientists at any given time become a source of resistance to new ideas. And our illustrations begin with the very origins of modern science. In his magisterial discussion of the Copernican revolution, Kuhn[3] tells us not only about the nonscientific opposition to the heliocentric theory but also about the resistance from the astronomer-scientists of the time. Even after the publication of *De Revolutionibus*, the belief of most astronomers in the stability of the earth was unshaken. The idea of the earth's motion was either ignored or dismissed as absurd. Even the great astronomer-observer Tycho Brahe remained a life-long opponent of Copernicanism; he was unable to break with the traditional patterns of thought about the earth's lack of motion. And his immense prestige helped to postpone the conversion of other astronomers to the new theory. Of course, religious, philosophical, and ideological conceptions were closely interwoven with substantive scientific theories in the culture of the scientists of that time, but it seems clear that the latter as well as the former played their part in the resistance to the Copernican discoveries.

Moving to the early nineteenth century, we learn that the scientists of the day resisted Thomas Young's wave theory of light because they were, as C. C. Gillispie says, faithful to a corpuscular model.[15] By the end of the century, when scientists had swung over to the wave theory, the validity of Young's earlier discovery was recognized. Substantive scientific theory was also one of the sources of resistance to Pasteur's discovery of the biological character of fermentation processes. The established theory that these processes are wholly chemical was held to by many scientists, including Licbig, for a long time.[16] The same preconceptions were also the source of the resistance to Lister's germ theory of disease, although in this case, as in that of Pasteur, various other factors were important.

Because it illustrates a variety of sources of scientific resistance to discovery, I shall return several times to the case of Mendel's theory of genetic inheritance. For the present, I mention it only in connection with the source of resistance under discussion, substantive scientific theories themselves. Mendelian theory, it seems clear, was resisted from the time of its announcement, in 1865, until the end of the century, because Mendel's conception of the separate inheritance of characteristics ran counter to the predominant conception of joint and total inheritance of biological characteristics.[17-18] It was not until botany changed its conceptions and concentrated its research on the separate inheritance of unit characteristics that Mendel's theory and Mendel himself were independently rediscovered by de Vries, a Dutchman, by Carl Correns, working in Tübingen, and by Erich Tschermak, a Viennese, all in the same year, 1900.

New conceptions about the electronic constitution of the atom were also resisted by scientists when fundamental discoveries in this field were being made at the end of the nineteenth century. The established scientific notion was that of the absolute physical irreducibility of the atom. When Arrhenius published his theory of electrolytic dissociation, his ideas met with resistance for a time, though eventually, thanks in part to Ostwald, the theory was accepted and Arrhenius was given the Nobel Prize for it.[19] Similarly, Lord Kelvin regarded the announcement of Röntgen's discovery of X-rays as a hoax, and as late as 1907 he was still resisting the discovery, by Ramsay and Soddy, that helium could be produced from radium, and resisting Rutherford's theory of the electronic composition of the atom, one of the fundamental discoveries of modern physics. Throughout his long and distinguished life in science Kelvin never discarded the concept that the atom is an indivisible unit.[20]

Let us take one final illustration, from contemporary science. In a recent case history of the role of chance in scientific discovery it was reported that two able scientists, who observed, independently and by chance, the phenomenon of floppiness in rabbits' ears after the injection of the enzyme papain, both missed making a discovery because they shared the established scientific view that cartilage is a relatively inert and uninteresting type of tissue.[21] Eventually one of the scientists did go on to make a discovery which altered the established view of cartilage, but for a long time even he had been blinded by his scientific preconceptions. This case is especially interesting because it shows how resistance occurs not only between two or more scientists but also within an individual scientist. Because of their substantive conceptions and theories, scientists sometimes miss discoveries that are literally right before their eyes.

## Methodological Conceptions

The methodological conceptions scientists entertain at any given time constitute a second cultural source of resistance to scientific discovery and are as important as substantive ideas in determining response to innovations. Some scientists, for example, tend to be antitheoretical, resisting, on that methodological ground, certain discoveries. "In Baconian science," says Gillispie, "the bird-watcher comes into his own while genius, ever theorizing in far places, is suspect. And this is why Bacon would have none of Kepler or Copernicus or Gilbert or anyone who would extend a few ideas or calculations into a system of the world."[15] Goethe too, as Helmholtz pointed out in his discussion of Goethe's scientific research, was antitheoretical.[22] A more recent discussion of Goethe's scientific work also finds him anti-analytical and anti-abstract.[15] Perhaps Helmholtz had been made aware of Goethe's antitheoretical bias because his own discovery of the conservation of energy had been resisted as being too theoretical, not sufficiently experimental. German physicists were probably antitheoretical in Helmholtz's day because they feared a revival of the speculations of the Hegelian "nature-philosophy" against which they had fought so long, and eventually successfully.

Viewed in another way, Goethe's antitheoretical bias took the form of a positive preference for scientific work based on intuition and the direct evidence of the senses. "We must look upon his theory of colour as a forlorn hope," says Helmholtz, "as a desparate attempt to rescue from the attacks of science the belief in the direct truth of our sensations."[22] Goethe felt passionately that Newton was wrong in analyzing color into its quantitative components by means of prisms and theories. Color, for him, was a qualitative essence projected onto the physical world by the innate biological character and functioning of the human being.

Later scientists also have resisted discovery because of their preference for the evidence of the senses. Otto Hahn, noted for his discoveries in radioactivity, who received the Nobel Prize for his splitting of the uranium atom in 1939, reports the following case:

> Emil Fischer was also one of those who found it difficult to grasp the fact that it is also possible by radioactive methods of measurement to detect, and to recognize from their chemical properties, substances in quantities quite beyond the world of the weighable; as is the case, for example, with the active deposits of radium, thorium, and actinium. At my inaugural lecture in the spring of 1907, Fischer declared that somehow he could not believe those things. For certain substances the most delicate test was afforded by the sense of smell and no more delicate test could be found than that![23]

Another methodological source of resistance is the tendency of scientists to think in terms of established models, indeed to reject propositions just because they cannot be put in the form of some model. This seems to have been a reason for resistance to discoveries in the theory of electromagnetism during the nineteenth century. Ampère's theory of magnetic currents, for example, was resisted by Joseph Henry and others because they did not see how it could be fitted into the Newtonian mechanical model.[24] They refused to accept Ampère's view that the atoms of the Newtonian model had electrical properties, which caused magnetic phenomena. And Lord Kelvin's resistance to Clerk Maxwell's electromagnetic theory of light was due, says Kelvin's biographer,[20] to the fact that Kelvin found himself unable to translate into a dynamic model the abstract equations of Maxwell's theory. Kelvin himself, in the lectures he had given in Baltimore in 1884, had said, "I never satisfy myself until I can make a mechanical model of a thing. If I can make a mechanical model I can understand it. As long as I cannot make a mechanical model all the way through I cannot understand; and that is why I cannot get the electromagnetic theory."[20] Thus, models, while usually extremely helpful in science, can also be a source of blindness.

Scientists' position on the usefulness of mathematics is a last methodological source of resistance to discovery. Some scientists are excessively partial to mathematics, other excessively hostile. Thus, when Faraday made his experimental discoveries on electromagnetism, Gillispie tells us, few mathematical physicists gave them any serious attention. The discoveries were regarded with indulgence or a touch of scorn as another example of the mathematical incapacity of the British, their barbarous emphasis on experiment, and their theoretical immaturity.[15] Clerk Maxwell, however, resolved that he "would be Faraday's mathematicus"—that is, put Faraday's experimental discoveries into more mathematical, general, and theoretical form. Initial resistance was thus overcome. Long ago Augustus De Morgan commented on the antimathematical prejudice of English astronomers of his time. In 1845, he pointed out, the Englishman John C. Adams had, on the basis of mathematical calculations, communicated his discovery of the planet Neptune to his English colleagues. Because they distrusted mathematics, his discovery was not published, and eight months later the Frenchman Leverrier announced and published his simultaneous discovery of the planet, once again on the basis of mathematical calculations. Because the French admired mathematics, Leverrier's discovery was published first, and thus he gained a priority over Adams.[25]

Mendel was another scientist whose ideas were resisted because of the antimathematical preconceptions of the botany of his time. "It must be admitted, however," says his biographer, H. Iltis,

> that the attention of most of the hearers [when he read his classic monograph, "Experiments in Plant-Hybridization," before the Brünn Society for the Study of Natural Science in 1865] was inclined to wander when the lecturer was engaged in rather difficult mathematical deductions; and probably not a soul among them really understood what Mendel was driving at. . . . Many of Mendel's auditors must have been repelled by the strange linking of botany with mathematics, which may have reminded some of the less expert among them of the mystical numbers of the Pythagoreans . . . [18]

Note that the alleged "difficult mathematical deductions" are what we should now consider very simple statistics. And it was not just the audience in Brünn that had no interest in or knowledge of mathematics. Mendel's other biographer, Krumbiegel, tells us that even the more sophisticated group of scientists at the Vienna Zoological-Botanical Society would have given Mendel's theory as poor a reception, and for the same reasons.

In some quarters the antimathematical prejudice persisted in biology for a long time after Mendel's discovery, indeed until after he had been rediscovered. In his biography of Galton, Karl Pearson reports that he sent a paper to the Royal Society in October 1900, eventually published in November 1901, containing statistics in application to a biological problem.[26] Before the paper was published, he says, "a resolution of the Council [of the Royal Society] was conveyed to me, requesting that in future papers mathematics should be kept apart from biological applications." As a result of this, Pearson wrote to Galton, "I want to ask your opinion about resigning my fellowship of the Royal Society." Galton advised against resigning, but he did help Pearson to found the journal *Biometrika*, so that there would be a place in which mathematics in biology would be explicity encouraged. Galton wrote an article for the first issue of the new journal, explaining the need for this new agency of "mutual encouragement and support" for mathematics in biology and saying that "a new science cannot depend on a welcome from the followers of the older ones, and [therefore] . . . it is advisable to establish a special Journal for Biometry".[27] It seems strange to us now that prejudice against mathematics should have been a source of resistance to innovation in biology only sixty years ago.

### Religious Ideas

Although we have heard more of the way in which religious forces outside science have hindered its progress, the religious ideas of scientists themselves constitute, after substantive and methodological conceptions, a third cultural source of resistance to scientific innovation. Such internal resistance goes back to the beginning of modern science. We have seen that the astronomer colleagues of Copernicus resisted his ideas in part because of their religious beliefs, and we know that Leibniz, for example, criticized Newton "for failing to make providential destiny part of physics."[15] Scientists themselves felt that science should justify God and his world. Gradually, of course, physics and religion were accommodated one to the other, certainly among scientists themselves. But all during the first half of the nineteenth century resistance to discovery in geology persisted among scientists for religious reasons. The difficulty, as Gillispie has put in on the basis of his classic analysis of geology during this period, "appears to be one of religion (in a crude sense) *in* science rather than one of religion *versus* scientists." The most embarrassing obstacles faced by the new sciences were cast up by the curious providential materialism of the scientists themselves.[5] When, in the 1840s, Robert Chambers published his *Vestiges of Creation*, declaring a developmental view of the universe, the theory of development was so at variance with the religious views that all scientists accepted that "they all spoke out: Herschel, Whewell, Forbes, Owen, Prichard, Huxley, Lyell, Sedgwick, Murchison, Buckland, Agassiz, Miller, and others".[5, p. 133; 28-29]

Religious resistance continued and was manifested against Darwin, of course, although many of the scientists who had resisted earlier versions of developmentalism accepted Darwin's evolutionary theory, Huxley being not the least among them. In England, Richard Owen offered the greatest resistance on scientific grounds, while in American and, in fact, internationally, Louis Agassiz was the leading critic of Darwinism on religious grounds.[5], [29-30]

In more recent times, biology, like physics before it, has been successfully accommodated to religious ideas, and religious convictions are no longer a source of resistance to innovation in these fields. Resistance to discoveries in the psychological and social sciences that stems from religious convictions is perhaps another story, but one that does not concern us here.

In addition to shared idea-systems, the patterns of social interaction among scientists also become sources of resistance to discovery. Here again we are dealing with elements that, on the whole, probably serve to advance science but occasionally produce negative, or dysfunctional, effects.

### Professional Standing

The first of these social sources of resistance is the relative professional standing of the discoverer. In general, higher professional standing in science is achieved by the more competent, those who have demonstrated their capacity for being creative in their own right and for judging the discoveries of others. But sometimes, when discoveries are made by scientists of lower standing, they are resisted by scientists of higher standing partly because of the authority the higher position provides. Huxley commented on this social source of resistance in a letter he wrote in 1852:

> For instance, I know that the paper I have just sent in is very original and of some importance, and I am equally sure that if it is referred to the judgment of my "particular" friend that it will not be published. He won't be able to say a word against it, but he will pooh-pooh it to a dead certainty. You will ask with wonderment, Why? Because for the last twenty years [. . . .] has been regarded as the great authority in these matters, and has had no one tread on his heels, until, at last, I think, he has come to look upon the Natural World as his special preserve, and "no poachers allowed," So I must manoeuvre a little to get my poor memoir kept out of his hands.[8, p. 367]

Niels Henrik Abel, early in the nineteenth century, made important discoveries on a classical mathematical problem, equations of the fifth degree.[31] Not only was Abel himself unknown but there was no one of any considerable professional standing in his own country, Norway (then part of Denmark), to sponsor his work. He sent his paper to various foreign mathematicians, the great Gauss among them. But Gauss merely filed the leaflet away unread, and it was found uncut after his death, among his papers. Ohm was another whose work, in this case experimental, was ignored partly because he was of low professional standing. The researches of an obscure teacher of mathematics at the Jesuit Gymnasium in Cologne made little impression upon the more noted scientists of the German universities.

Perhaps the classical instance of low professional standing helping to create resistance to a scientist's discoveries is that of Mendel. The notion that Mendel was "obscure," in the sense that his work did not come to the attention of competent and noted professionals in his field, can no longer be accepted. First of all, the proceedings volume of the Brünn Society in which his monograph was printed was exchanged with proceedings volumes of more than 120 other societies, universities, and academies at home and abroad. Copies of his monograph went to Vienna and Berlin, to London and Petersburg, to Rome and Uppsala.[18] In London, according to Bateson, the monograph was received by the Royal Society and the Linnaean Society.[32] Moreover, we know from the extensive correspondence between them—correspondence which was later published by Mendel's rediscoverer, Correns—that Mendel sent his paper to one of the distinguished botanists of his time, Carl von Nägeli of Munich.[15, 17, 18] Von Nägeli resisted Mendel's theories for a number of reasons: because his own substantive theories about inheritance were different and because he was unsympathetic to Mendel's use of mathematics, but also because he looked down, from his position of authority, upon the unimportant monk from Brünn. Mendel had written deferentially to von Nägeli, in letters that amounted to small monographs. In these letters, Mendel addressed von Nägeli most respectfully, as an acknowledged master of the subject in which they were both interested. But von Nägeli was the victim of his own position as a scientific pundit. Mendel seemed to him a mere amateur expressing fantastic notions, or at least notions contrary to his own. Von Nägeli's letters to Mendel seem unduly critical to present readers, more than a little supercilious. Nevertheless, the modest Mendel was delighted that the great man had even deigned to reply and sent cordial thanks for the gift of von Nägeli's monograph. On both sides, von Nägeli was defined as the great authority, Mendel as the inferior asking for consideration his position did not warrant. Ironically, Mendel took von Nägeli's advice, to change from experiments on peas to work on hawkweed, a plant not at all suitable at that time for the study of inheritance of separate characteristics. The result was that Mendel labored in a blind alley for the rest of his scientific life.

Nor was von Nägeli unique. Others, such as W. O. Focke, Hermann Hoffman, and Kerner Von Marilaun, also dismissed Mendel's work because he seemed "an insignificant provincial" to them. Focke did list Mendel's monograph in his own treatise, *Die Pflanzenmischlinge*, but only for the sake of completeness. Focke paid much more attention to those botanists who had produced quantitatively large and apparently more important contributions—

men such as Kölreuter, Gärtner, Wichura, and Wiegmann, of higher professional standing.[33] Certainly, in this case, quantity of publication was inadequate as a measure of professional worth. Focke's listing of Mendel served only to bring his work, directly and indirectly, to the attention of Correns, de Vries, and von Tschermak after they had independently rediscovered the Mendelian principle of inheritance.

Mendel met with resistance from the authorities in his field after his discovery was published. But sometimes men of higher professional standing sit in judgment on lesser figures *before* publication and prevent a discovery's getting into print. This can be illustrated by an incident in the life of Lord Rayleigh. For the British Association meeting at Birmingham in 1886, Rayleigh submitted a paper under the title, "An Experiment to Show that a Divided Electric Current May Be Greater in Both Branches Than in the Mains." "His name," says his son and biographer,

> was either omitted or accidentally detached, and the Committee "turned it down" as the work of one of those curious persons called paradoxers. However, when the authorship was discovered, the paper was found to have merits after all. It would seem that even in the late 19th century, and in spite of all that had been written by the apostles of free discussion, authority could prevail when argument had failed![34]

So says the fourth Baron Rayleigh, and we may wonder whether his remark does not still apply, some seventy-five years later.

### Professional Specialization

Another social source of resistance is the pattern of specialization that prevails in science at any given time. On the whole, of course, as with any social or other type of system, such specialization is efficient for internal and environmental purposes. Specialization concentrates and focuses the requisite knowledge and skill where they are needed. But occasionally the negative aspect of specialization shows itself, and innovative "outsiders" to a field of specialization are resisted by the "insiders." Thus, when Helmholtz announced his theory of the conservation of energy, it met with resistance partly because he was not a specialist in what we now think of as physics. Referring in the later years of his life to the opposition of the achnowledged experts, Helmholtz said he met with such a remark as this from some of the older men: "This has already been well known to us; what does this young medical man imagine when he thinks it necessary to explain so minutely all this to us?" ([8], p. 97). To be sure, on the other side, medical specialists have a long history of resisting scientific innovations from what they define as "the outside." Pasteur met with violent resistance from the medical men of his time when he advanced his germ theory. He regretted that he was not a medical specialist, for the medical men thought of him as a mere chemist poaching on their scientific preserves, not worthy of their attention. In France, even before Pasteur, François Magendie had met with resistance for attempting to introduce chemistry into medicine.[35] If medicine now listens more respectfully to nonmedical science and its discoveries, it is partly because many nonmedical scientists have themselves become experts in a variety of medical-science specialties and so are no longer "outsiders."

## Societies, "Schools," and Seniority

Scientific organizations, as we may safely infer from their large number and their historical persistence, serve a variety of useful purposes for their members. And, of course, scientific publications are indispensable for communication in science. But occasionally, when organizations or publications are incompetently staffed and run, they may serve as another social source of resistance to innovation in science. There have been no scholarly investigations into the true history of our scientific organizations and publications, but something is known and points in the direction I have suggested. In the early nineteenth century, for example, even the Royal Society fell on bad days. Lyons tells us that a contemporary, Granville, "severely criticized the shortcomings of the Society" during that period.[36] Granville gave numerous instances in which the selection or rejection of papers by the Committee of Papers was the result of bad judgment. Sometimes the paper had not been read by any Fellow who was an authority on the subject with which it dealt. In other cases, none of the members of the committee who made the judgment could have had any expert opinion in the matter. It was such an incompetent committee, for example, that resisted Waterston's new molecular theory of gases when he submitted a paper making this contribution. The referee of the Royal Society who rejected the paper wrote on it, "The paper is nothing but nonsense." As a result, Waterston's work lay in utter oblivion until rescued by Rayleigh some forty-five years later, (*12,* p. 26). Many present-day misjudgments of this kind probably occur, although the multiplicity of publication outlets now provides more than one chance for a significant paper ignored by the incompetent to appear in print.

The rivalries of what are called "schools" are frequently alleged to be another social source of resistance in science. T. H. Huxley, for example, is reported to have said, two years before his death, "'Authorities,' 'disciples,' and 'schools' are the curse of science; and do more to interfere with the work of the scientific spirit than all its enemies."[37] Murray suggests that the supposed warfare between science and theology is equaled only by the warfare among rival schools in each of the scientific specialties. Unfortunately, just what the term "school" means is usually left unclear, and no empirical evidence of anything but the most meager and unsystematic character is ever offered by way of illustration.[38] No doubt some harmful resistance to discovery, as well as some useful competition, comes out of the rivalry of "schools" in science, but until the concept itself is clarified, with definite indicators specified, and until research is carried out on this more adequate basis, we can only feel that "there is something there" that deserves a scholarly treatment it has not yet received.

That the older resist the younger in science is another pattern that has often been noted by scientists themselves and by those who study science as a social phenomenon. "I do not," said Lavoisier in the closing sentences of his memoir *Reflections on Phlogiston* (read before the Academy of Sciences in 1785), "expect my ideas to be adopted all at once. The human mind gets creased into a way of seeing things. Those who have envisaged nature according to a certain point of view during much of their career, rise only with difficulty to new ideas. It is the passage of time, therefore, which must confirm or destroy the opinions I have presented. Meanwhile, I observe with great satisfaction that the young people are beginning to study the science without prejudice. . . ."[15] Or again, Hans Zinsser remarks in his autobiography, "That academies and learned societies—commonly dominated by the older foofoos of any profession—are slow to react to new ideas is in the nature of things. For, as Bacon says, *scientia inflat*, and the dignitaries who hold high honors for past accomplishment do not usually like to see the current of progress rush too rapidly out of their reach[39]."

Now, of course, the older workers in science do not always resist the younger in their innovations, nor can it be physical aging in itself that is the source of such resistance as does occur. If we scrutinize carefully the two comments I have just quoted and examine other, similar ones with equal care, we can see that *aging* is an omnibus term that actually covers a variety of cultural and social sources of resistance. Indeed, we may put it this way, that as scientists get older they are more likely to be subject to one or another of the several cultural and social sources of resistance I have analyzed here. As a scientist gets older, he is more likely to be restricted in his response to innovation by his substantive and methodological preconceptions and by his other cultural accumulations; he is more likely to have high professional standing, to have specialized interests, to be a member or official of an established organization, and to be associated with a "school." The likelihood of all these things increases with the passage of time, and so the older scientist, just by living longer, is more likely to acquire a cultural and social incubus. But this is not always so, and the older workers in science are often the most ardent champions of innovation.

After this long recital of the cultural and social sources of resistance, by scientists, to scientific discovery, I need to emphasize a point I have already made. That some resistance occurs, that it has specifiable sources in culture and social interaction, that it may be in some measure inevitable, is not proof either that there is more resistance than acceptance in science or that scientists are no more open-minded than other men. On the contrary, the powerful norm of open-mindedness is science, the objective tests by which concepts and theories often can be validated, and the social mechanisms for ensuring competition among ideas new and old—all these make up a social system in which objectivity is greater than it is in other social areas, resistance less. The development of modern science demonstrates this ever so clearly. Nevertheless, some resistance remains, and it is this we seek to understand and thus perhaps to reduce. If "the edge of objectivity" in science, as Charles Gillispie has recently pointed out, requires us to take physical and biological nature as it is, without projecting our wishes upon it, so also we have to take man's social nature, or his behavior in society, as it is. As persons in society, scientists are sometimes the agents, sometimes the objects, of resistance to their own discoveries.[40]

## References and Notes

1. S. C. Gilfillan, *The Sociology of Invention* (Chicago, Ill.: Follet, 1935); B. Barber, *Science and the Social Order* (Glencoe, Ill.: Free Press, 1952), chap. 9.
2. P. G. Frank, in *The Validation of Scientific Theories*, P. G. Frank ed. (Boston, Mass.: Beacon Press, 1957); J. Rossman, *The Psychology of the Inventor* (Washington, D.C.: Inventors Publishing Co., 1931), chap. 11; R. H. Shryock, *The Development of Modern Medicine* (Philadelphia, PA.: University of Pennsylvania Press, 1936), chap. 3; B. J. Stern, in *Technological Trends and National Policy* (Washington, D.C.: Government Printing Office, 1937); V. H. Whitney, *Am. J. Sociol.* 56 (1950) 247; J. Stamp, *The Science of Social Adjustment* (London, Macmillan, 1937), pp. 34ff.; A. C. Ivy, *Science* 108 (1948): 1.
3. T. S. Kuhn, *The Copernican Revolution* (Cambridge, Mass.: Harvard University Press, 1957).
4. A. N. Whitehead, *Science and the Modern World* (New York: Macmillan, 1947), chap. 1; R. K. Merton, *Osiris* 4, pt. 2 (1938).
5. C. C. Gillispie, *Genesis and Geology* (Cambridge, Mass.: Harvard University Press, 1951).
6. R. Oppenheimer, *The Open Mind* (New York: Simon and Schuster, 1955).
7. Quoted from Helmholtz's *Vorträge und Reden* in R. H. Murray (8).
8. R. H. Murray, *Science and Scientists in the Nineteenth Century* (London: Sheldon, 1825).
9. Lord Kelvin also commented on the "resistance" to Faraday. In his article on "Heat" for the 9th edition of the *Encyclopaedia Britannica* he made a comment on the circumstance "that fifty years passed before the scientific world was converted by the experiments of Davy and Rumford to the rational conclusion as to the non-materiality of heat: 'a remarkable instance of the tremendous inefficiency of bad logic in confounding public opinion and obstructing true philosophic thought.'" [S. P. Thompson, *The Life of William Thomson. Baron Kelvin of Largs* (London; Macmillan, 1910)].

10. M. Planck, *Scientific Autobiography*, F. Gaynor, trans. (New York: Philosophical Library, 1949).
11. See D. L. Watson, *Scientists Are Human* (London: Watts, 1938).
12. W. Trotter, *Collected Papers* (London: Humphrey Milford, 1941).
13. W. I. B. Beveridge, *The Art of Scientific Investigation* (New York: Random House, rev. ed., 1959).
14. H. Levy, *Universe of Science* (New York: Century, 1933), p. 197.
15. C. C. Gillispie, *The Edge of Objectivity* (Princeton, N.J.: Princeton University Press, 1960).
16. R. Vallery-Radot, *The Life of Pasteur*, R. L. Devonshire, trans. (New York: Garden City Publishing Co., 1926), pp. 175, 215.
17. I. Krumbiegel, *Gregor Mendel und das Schicksal Seiner Entdeckung* (Stuttgart: Wissenschaftliche Verlagsgesellschaft, 1957).
18. H. Iltis, *Life of Mendel*, E. Paul and C. Paul, trans. (New York: W. W. Norton, 1932).
19. J. J. Thomson, *Recollections and Reflections* (London: Bell, 1936), p. 390.
20. S. P. Thompson, *The Life of William Thomson: Baron Kelvin of Largs* (London: Macmillan, 1910).
21. B. Barber and R. C. Fox, *Am. J. Sociol.* 64 (1958): 128.
22. H. von Helmholtz, *Popular Scientific Lectures* (New York: Appleton, 1873).
23. O. Hahn, *New Atoms, Progress and Some Memories* (New York: Elsevier, 1950), pp. 154–55.
24. T. Coulson, *Joseph Henry: His Life and Work* (Princeton, N.J.: Princeton University Press, 1950), p. 36.
25. S. E. De Morgan, *Memoir of Augustus De Morgan* (London: Longmans, Green, 1882).
26. K. Pearson, *The Life, Letters and Labours of Francis Galton* (Cambridge, England: Cambridge University Press, 1924), vol. 3, pp. 100, 282–83.
27. *Biometrika* (1901–2): 7.
28. That scientists were religious also, and in the same way, in America can be seen in A. H. Dupree (*29*).
29. A. H. Dupree, *Asa Gray* (Cambridge, Mass.: Harvard University Press, 1959).
30. E. Lurie, *Louis Agassiz: A Life in Science* (Chicago, Ill.: University of Chicago Press, 1960).
31. O. Ore, *Niels Henrik Abel: Mathematician Extraordinary* (Minneapolis, Minn.: University of Minnesota Press, 1957).
32. R. A. Fisher, *Ann. Sci.* 1 (1933): 116.
33. H. F. Roberts, *Plant Hybridization before Mendel* (Princeton, N.J.: Princeton University Press, 1929), pp. 210–211.
34. R. J. Strutt, *John William Strutt, Third Baron Rayleigh* (London: Arnold, 1924), p. 228.
35. J. M. D. Olmstead, *François Magendie, Pioneer in Experimental Physiology and Scientific Medicine in the 19th Century* (New York: Schuman, 1944), pp. 173–75.
36. H. Lyons, *The Royal Society, 1661–1940* (Cambridge, England: Cambridge University Press, 1944), p. 254.
37. C. Bibby, *T. H. Huxley: Scientist, Humanist, and Educator* (New York: Horizon, 1959), p. 18.
38. For the best available sociological essay, see F. Znaniecki, *The Social Role of the Man of Knowledge* (New York: Columbia University Press, 1940), chap. 3.
39. H. Zinsser, *As I Remember Him: The Biography of R.S.* (Boston, Mass.: Little,

Brown, 1940), p. 105.

40. For invaluable aid in the preparation of this article I am indebted to Dr. Elinor G. Barber. The Council for Atomic Age Studies of Columbia University assisted with a grant for typing expenses.

# 6

# The Functions and Dysfunctions of "Fashion" in Science: A Case for the Study of Social Change

About a dozen years ago, in a paper dealing with "fashion" in women's clothes, a colleague and I made the following statement:

> In social science usage, "fashion" is still an overgeneralized term. One writer lists the following "fields of fashion": values in the pictorial arts, architecture, philosophies, religion, ethical behavior, dress, and the physical, biological, and social sciences. "Fashion" has also been used in reference to language usages, literature, food, dance music, recreation, indeed, the whole range of social and cultural elements. The core of meaning in the term for all these different things is "changeful," but it is unlikely that the structures of behavior in these different social areas and the consequent dynamics of their change are all identical. "Fashion," like "crime," has too many referents; it covers significantly different kinds of social behavior.[1]

Unfortunately, today there is still lacking a satisfactory social science analysis of what those who work in the physical, biological, and behavioral sciences often call "fashion."[2]

Besides this fault of overgeneralization by social scientists,, there are other shortcomings in the typical references to "fashion" that are made by the working scientists themselves, who do not claim, of course, to be giving a systematic analysis. The first of these shortcomings is to use "fashion" in a quite commonsense way, as a mere label which begs the analysis that is required. An example of this can be found in the response of a scientist to some questions asked him in a study of the flow of scientific information among scientists.[3] In connection with a question about the ways in which some scientific work published earlier is revived by someone and then communicated widely, the scientist-respondent refers to:

> . . . some work done and published in 1942 (which) just came back three years ago in a symposium on lipids. It had been published in 1942 in a German biochemical journal. Then, just three years ago, someone in California, working on lipid separation, used this material. . . . Now it has been used greatly. . . . (Why had this material not been used in the intervening years?) I don't know. Guess it's just science following fashion.

As Herbert Menzel properly comments, "That 'science follows fashion' gives a name to the problem, but does not account for it."[4]

A second shortcoming, which reveals both the commonsense approach and the negative view of "fashion," is the tendency to exaggeration of the extent to which "fashion" in science occurs. One clue to this exaggeration is the failure to give any evidence for the alleged rampancy of "fashion." The following example is taken from a signed editorial in *Science* by Professor Ernst Mayr of Harvard, the distinguished biologist: "There has long been a bandwagon tendency in American science," he says, "but today it seems particularly rampant. This seems true of the physical sciences and particularly of the biological sciences."[5]

A third shortcoming of the use of the term by working scientists is that any reference they may make to psychological, social, or cultural factors tends to be in terms of imprecise, nontechnical, unsystematic notions such as "human nature," or "the climate of the times," or being "in tune with the times." An example is provided by the following statement from an editorial in *Science* criticizing the United States government's program of support for scientific research:

> Still another negative feature is a psychological one. Scientists, like other human beings, are affected by fads. They tend to go with the crowd. The research worker who does not go with the crowd encounters a rather bleak climate. He is likely to be regarded by administrators and laymen as an odd fellow who is not in tune with the times. Under this pressure, undue emphasis develops on glamorous areas."[6]

Finally, there is a shortcoming in the usage of "fashion" by working scientists, which is nowadays, and for obvious reasons, more likely to occur in the behavioral than the natural sciences. This shortcoming is the tendency to use the term "fashion" with its negative implications, as an ideological stick with which to beat some field of research in which there has been a recent increase and which the user does not like, or likes less than some other field. Thus, a few years ago, when the increase in small-group research in sociology and social psychology was near the peak of the large increase it had had in the preceding ten years, one social scientist, who preferred what he considered to be the "big" problems of economic and political behavior, tried to explain, and perhaps thus "explain away," the "fashion" for small-group research as due to the political fears American social scientists had for dealing with these "big" problems. The microsociology of small-group research was thus ideologically criticized as being a poor substitute for the necessary macrosociology that this critic preferred. No evidence was offered that in fact American social scientists were not dealing in considerable measure with macrosociological problems, and no attention was paid to some of the other social and cultural sources of small-group research besides the ideological one imputed to it. As we shall suggest later on, these other sources have probably played a large part in the increase of small-group research during the last fifteen years.

So much for the inadequacy of present references by both social scientists and working scientists in all fields to "fashion" in science. Now we can perhaps make our way toward a more satisfactory social science explanation by clarifying the following several matters: (1) What is the essential character or element of "fashion" in science? (2) What are some of the persisting social and cultural sources of this element of "fashion"? (3) Given these sources of "fashion" in science, and given some other norms and conditions for successful science, what are some of the functions and dysfunctions of "fashion" in science? (4) Finally, what is the patterned response of scientists to "fashion" and can this pattern be explained in terms of the functions and dysfunctions of "fashion" in their field?

## Clarifying Concepts of "Fashion" in Science

### 1. The Essential Element of "Fashion" in Science

Just as is the case with the usage of "fashion" in its most general sense, so the essential element in its usage with regard to science is "changefulness." The working scientist who refers to some particular "fashion" in science or to the widespread occurrence of "fashion" throughout science is always at least pointing to some change that has occurred, and often being somewhat critical of that change as well. Since science is full of changes, there is always something to cry "fashion" about.[7] Unfortunately, however, such cries are usually not specific about some important dimensions of change, such as the type and the rate of change. The term "fashion" in itself does not specify whether the change is from one type of basic scientific specialty to another basic type, as from physics to biology, or between two fairly closely related scientific specialties, or types, as from one branch of nuclear physics to another, or from one technique in one branch of nuclear physics to another technique in the same branch. Science as a whole is large, its specialties are more and less closely related, and discussion of changes or "fashions" should attempt to specify differences in the degree of relatedness of older and newer types or fields of work. Nor does the term "fashion" in itself say anything precise about the rate of change, which may differ considerably, whether between closely related specialties or basically different ones. Rates of change could be specified in terms of such indicators as shifts of scientific personnel, increase or decrease in the number of publications, the opening up or closing down of professorships and other research positions in a specialty, and perhaps even in the character of theoretical or methodological alterations. Discussions of "fashion" or change should also allow for the fact that what one observer sees as a certain rate of change may be seen quite differently, that is, as faster or slower, by other scientists, whose positions on the social and cultural structure of science differ from his. For example, scientists who are high in the social structure of prestige in their field may see some change in quite a different way from those who are lower in that structure, expecially if the change involves an alteration in their relative prestige positions. Or, scientists brought up with the older theoretical and methodological ideas may see change as proceeding much faster than newcomers to the field see it. Discussions of "fashion" in science should attempt to specify rates of change both as they appear to the objective observer and as they appear to participants differently located in the structure and culture of science. Change, then, is a constant element in science, but it is not a simple or homogeneous phenomenon.

### 2. Persisting Social and Cultural Sources of "Fashion" or Change in Science

Because change is a constant in science, and because it has a variety of persisting social and cultural sources, a more satisfactory social science analysis of "fashion" in science requires an examination of these sources. Among its several advantages, such an examination will open up the possibility of constructing better indicators of those various types and rates of change in science which vague cries of "fashion" cannot discriminate. Before discussing these several sources of change and providing some illustration for each of them, several points about these sources considered collectively should be noted. First, the sources mentioned here are important, but they constitute neither an exhaustive nor a systematic list. Second, the sources are not discussed in any necessary order of relative importance. Third, we must remember especially that any one of these sources, but usually more than one in some combination, will be the determinant of change in science. Finally, under different conditions, both the relative importance of each source and the particular combination of sources that brings about change, will be different.

*New ideas or concepts.* New concepts are obviously one important source of change in science. Indeed, T. S. Kuhn says it is of the essence of a scientific revolution that there be a change from one set of concepts, one model, one paradigm, as he calls it, to another.[8] The history of science is full of "fashions" started by new ideas. To mention two chosen nearly at random, from a paper by Holton, it was some new ideas about molecular beams developed by Professor I. I. Rabi, after studying with Otto Stern in Hamburg, that led "soon after, both in independent laboratories as well as in those of Rabi and his associates" to a great deal of new work both in this field and in "neighboring parts of the same field."[9] Holton continues, "The excitement of this field as a whole and its fruitfulness are attested by the large rate of inflow of new persons, including many outstanding experimental and theoretical physicists."[10] In short, what some might call a new "fashion" occurred. Another, earlier example mentioned by Holton was the "fashion" created by the new concepts about magnetic fields around wires that carry direct current, discovered around 1820 by Oersted, Biot, Savart, and Ampere.[11] In recent social science, of course, we have the example of the "fashion" or change referred to above, in small-group research, that was brought about in considerable measure because of some new concepts about the structure and functions of these groups formulated by R. F. Bales.[12]

*New methods.* As well as new ideas, new methods for research are important in bringing about change in science. Indeed, in two of the cases just mentioned, that of Rabi's work on molecular beams and Bales's work on small groups, it was the concurrence of new methods and new ideas for studying these phenomena that made them so attractive and "fashionable." Of course, new methods or tools for research may occur somewhat independently of basic new scientific concepts, as in the cases of the telescope or the electron microscope. That is to say, relatively-more-empirical technology sometimes develops in ways that bring about important change in theoretical science.

*New access.* New or improved accessibility to the necessary data is another source of change or "fashion" in science. Easier access to both live and dead human bodies in the nineteenth century, for example, was of great benefit to the human biological sciences.[13] In this case, easier access resulted from a change in public norms and attitudes. In other cases, easier access results from the development of new instruments of research. Thus, there is currently a great change or "fashion" in "the space sciences" because rocket and space technology bring hitherto inaccessible areas of the universe within the reach of scientists. As we shall have occasion to note again, later, the current "fashion" in medical sociology is owing partly to the new access to hospitals and patients given by medical doctors to sociologists.

*New recruits.* An essential component of any change or "fashion" in science is, of course, new recruits to the new ideas, new methods, or newly accessible data. There are two types of new recruits. One consists of the novices in the relevant scientific specialties, especially those who are more or less anxiously looking for subjects for doctoral dissertations and who find rich and rewarding opportunities in the new line of work. Another important type consists of the older man in the field, often those who are competent but not particularly creative men, but also sometimes even the most creative men who recognize outstandingly good new idea or method. Holton, it will be remembered, said that "many outstanding experimental and theoretical physicists" were attracted by the new ideas and methods discovered by Rabi for the study of molecular beams.[14] For the study of different changes or "fashions" in science, it would be interesting to know the different proportions of these two types of new recruits for each change.

*New funds.* An increase in available funds for research is often one of a combination of sources of change or "fashion" in science. Here again, a nearly random choice of two recent examples from among the very large number that could be given will perhaps suffice. In his treatise on the diffusion of innovations, Rogers points out that since the mid-1950s there has been a proliferation of researches using sociometric methods on how agricultural innovations diffuse among farmers. The practical importance of the knowledge gained thereby and the increase in private and public funds for research have been the causes of this proliferation. "Most of these studies," says Rogers, "have been financed by state agricultural experiment stations or the USDA (but also in very recent years by agricultural companies). Federal and state agencies spend sizable sums for research on agricultural technology. Their administrators have been convinced of the value of sociological inquiry to trace the diffusion of these research results to farm people."[15] A similar case is found in the recent "fashionable" development of oceanography. "In behalf of science and education legislation," says John Walsh, "the cold war argument was often employed. A clear example of where it worked is in oceanographic research. Ten years ago the annual federal budget for oceanography was $10 million. By 1961 it had risen to $62 million, and in President Kennedy's first budget the following year it soared to $103 million. Another spurt took it to $123 million in fiscal 1963."[16]

*New professorships and other positions.* The availability of new university chairs and other scientific research positions is another source of change or "fashion" in science. An example of this has been described in excellent historical detail for nineteenth-century German science (with special emphasis on physiology) by Ben-David and Zloczower.[17] In the early decades of the nineteenth century, work in physiology at the German universities was "sporadic and haphazard." By 1828 physiology as an experimental discipline was represented in only six German universities by seven lecturers, no professors. Gradually, competence in physiology came to be a prerequisite for attaining the established chairs in anatomy. Some chairs of anatomy were filled by physiologists, but where older men survived, new chairs for physiology were established independently. Finally, anatomy and physiology were entirely separated. As a result, during the 1850s and 1860s many new chairs in physiology were established and filled by younger men with special training and research ability in that specialty. "Between 1855 and 1874 twenty-six scientists were given their first appointment to chairs of physiology. . . . Ten of these were appointed between 1855–59 alone. But therewith the discipline reached the limit of its expansion in the German university system." The number of chairs for physiology alone, which was 19 in 1873, was only 20 in 1880, in 1890, and in 1900. "Between 1875 and 1894, only nine scholars received appointments to chairs in physiology, stepping into chairs vacated by their incumbents." Aspiration to a professorship in physiology during the late 1870s and 1880s was "all but hopeless" because the cohort of first incumbents lived long and held their tenure for thirty and forty years or more. "Du Bois Reymond monopolized the chair in Berlin from 1858–96; Brücke reigned in Vienna for four decades, 1849–90; Echard held the chair in Giessen from 1855–91." The earlier "fashion" for physiology went out. Or, as Ben-David and Zloczower put it, "The result was that research in physiology lost momentum. A count of discoveries relevant to physiology in Germany shows that 321 such discoveries were made during the twenty years period of rapid expansion between 1855–74 compared with 232 during the subsequent (and 168 during the preceding) twenty years." Since there now were better ways of attaining a professorship than through physiology, new "fashions" arose for other fields. Hygiene professorships increased from one in 1873 to 19 in 1900. "Psychiatry grew from one chair to 16 and ophthalmology from 6 to 21 during the same period, while pathology, which had only 7 chairs in 1864 had reached 18 by 1880. The enthusiasm for physiology cooled considerably." Thus are "fashions" created in part by the availability of new professorships and other positions. The same process has occurred in many of the fields of the natural and social sciences in American universities during the last fifty to seventy-five years.

*Social problems.* One last source of change or "fashion" in science, a source that may exert its influence directly, but more often works indirectly, or in combination with other sources, for example, with the provision of new funds for research, consists of the "social problems" of the time. "Social problems" are those types of behavior which many people in the society have come to define as both undesirable and remediable. Mental illness, juvenile delinquency, cancer, have come to be defined as "social problems" in which the help of science is required. Felt "social problems," like these and others, produce support for various scientific specialties, support which is gladly taken up by some scientists but defined as merely the maker of "fashion" by others.

### 3. Functions and Dysfunctions of "Fashion" in Science

Given these several sources of change or "fashion" in science, if we consider some of the consequences they have, consider them especially in the light of certain important goals and norms of science, we can provide an answer to our third question, What are the functions and dysfunctions of "fashion" in science? Now because many times the terms of "fashion" carry strongly negative implications, we shall discuss dysfunctions first. We should also note, before we begin, an important point to which we shall return later, namely, that the same change or "fashion" may have mixed consequences, functional for some goals and norms of science, dysfunctional for others.

*Dysfunctions.* The primary goal of science is discovery of new ways of understanding the physical, biological, and social worlds. Essential to the support of this goal are the norms of science, which place a high value on the originality and autonomy of the individual scientist. Therefore, "fashion" is dysfunctional for science insofar as it involves a failure to maintain the norms of originality and autonomy, or at least not to achieve them in the fullest measure. Another dysfunction of "fashion" in science is that it sometimes results in what those who cry "fashion" think is an improper distribution of talents, efforts, and funds among the various scientific specialties. It may be the more imaginative men, as Ernst Mayr says,[18] who go into new fields, "glamorous" fields, as those critical of a "fashion" call them, where the most funds are, leaving the more orthodox men behind without sufficient talent to exploit all the opportunities for discovery that remain in the older field. Incidentally, notice the apparent contradiction between this assertion that it is the more imaginative men who follow "fashion" and the previous point that it is the less original men. We shall resolve this apparent contradiction in a moment. These are two possible *direct* dysfunctional consequences of "fashion" in science. A *derived* dysfunctional consequence may be that some of the young men who go into a field that is "fashionable" when they are narrowly trained for that special field and may become obsolete as scientists when that field is worked out and new "fashions" emerge.

*Functions.* However, "fashion" in science has its functions, too. First, even if those who follow others into a "fashionable" field are less than completely original and autonomous, they may be showing more originality and autonomy in recognizing a good new idea and pursuing it than in staying with some older, unprofitable line of thought. We see why the contradiction we mentioned is only "apparent." It is utopian to expect all scientists to be continuously and highly original. Second, the shifting of men, funds, and professorships into a newly "fashionable" field has the function, often, of getting a great deal of useful and necessary work done in that new field. Third, and finally, the excitement generated by a new and "fashionable" idea contributes to the morale of scientists, especially those who like to feel they are in the vanguard of the group that is struggling for victory over the unknown. Science is not without its own deadening routines, its own needs for coming a little nearer, once in a while at least, to its primary goal of discovery.

*The problem of a functional calculus.* Our earlier statement, that the same change or "fashion" may have mixed consequences, both dysfunctional and functional, is now perhaps a little clearer. Following a "fashion" may not express the highest originality, but it may express a lesser and functionally necessary originality still. Or it may involve the dysfunctional movement away from fields that retain some fertility, but the functional source of the movement is the appearance of even more fertile opportunities. If we had some functional calculus by which we could always make some swift, certain, and precise weighing up of functional and dysfunctional consequences, if we could always establish the fact of a net advantage or disadvantage for science, we would have more understanding and control over the changes or "fashions," which must inevitably occur in every area of science. But since we do not yet have such a functional calculus, scientists must often have mixed feelings toward particular changes or "fashions" in their own fields or in those that affect their own fields.

## 4. Ambivalence as the Patterned Response of Scientists to "Fashion" in Science

Now at long last, we can see why ambivalence is the usual and socially patterned response of those who speak of "fashion" in science.[19] Those who see only good in some change in science are not likely to speak of "fashion" at all. They are more likely to refer to the changing field as a "hot" field, and the "hotter" the field the better it is. But those who see some dysfunctional consequences, who dislike some of what they see a change bring about, are likely to use the term "fashion" because of its negative implications for science. Still, even these scientists usually cannot ignore the fact that changes or "fashions" in science usually have some positive or functional consequences as well. Note the following examples of ambivalence expressed by two distinguished scientists:

It is both inevitable and good that the dazzling achievements of molecular genetics have attracted wide attention. It is probably also inevitable, but not so good, that a bandwagon effect [that is, "fashion"] had led some people—not only immature students but some scientists who should have known better—to proclaim that molecular genetics is all that there is or should be to genetics. Genetics and biology must, however, deal not with one but with several levels of biological integration [Dobzhansky].[20]

A massive follow-up of new discoveries is normally highly productive, and no damage would be done if it were not for the fact that abandoned fields are rarely exhausted. When talent is diverted from them, science suffers an irreparable loss. . . . [We are] justified in fostering exploitation of breakthroughs, but it seems unwise . . . to pour most . . . funds into the glamor fields . . . the new should supplement the classical and not totally displace it [Mayr].[21]

It is my impression that the older, more established scientists are more likely than the younger ones to speak of "fashion," but for this impression I have no systematic evidence. In any case, the ambivalence of scientists toward "fashion" in science, is a socially structured ambivalence, structured by the fact that changes in science, usually have both recognizably functional and dysfunctional consequences for the goals and norms of science.

*Conclusion*

Perhaps we can now say that we have a better general understanding of "fashion" in science. And it seems to me that the essence of what this better understanding tells us is that we should give up the usage of the sociologically vague and morally invidious term "fashion" for the field of science and always speak instead of what is really at issue here, namely, the types, the rates, the sources, the variously functional and dysfunctional consequences of change, and the patterned responses of scientists to that change.

## Some Notes on the "Fashionableness" of Medical Sociology, 1945 Onward

Now let us look at recent changes in the development of medical sociology, changes that might be defined by some as a "fashion," and see how our general analysis applies. I hope that these will be considered only "notes" on the subject, since I have not undertaken the new research that would be necessary for a more satisfactory discussion. I have used what data I could find, and where they are lacking, I venture my own unsupported impressions. Still, I think something useful will emerge if we look at the recent "fashionableness" of medical sociology in terms of rates and types of change, sources of change, and some functions and dysfunctions of change.

## 1. The Rate of Change

We can take as our baseline for calculating the rate of change in medical sociology a paper published in 1951 in the *American Sociological Review*, the official journal of the American Sociological Association, by Professor Oswald Hall of McGill University, who was a pioneer among pioneers in this field.[22] It was intended as a defining and justificatory survey of the field of medical sociology. But its contents contain no mention of the numbers of research or teachers in the field, nor is the term "medical sociology" ever used. Hall, who had himself worked on a dissertation, "The Informal Organization of Medical Practice," during the late 1930s, which was accepted at the University of Chicago in 1944, mentions only four research works in medical sociology, two of them his own papers taken from his dissertation. Now Hall certainly knew of more works in the field of medical sociology than these; indeed he refers to them in his dissertation. But it is striking that his public view of the field, and the view of others who were in any way acquainted with it, I think, was a view that saw only a very little developed field of sociological specialization.

From 1951 on, however, medical sociology grows fast and probably at an accelerating rate. The first of two important surveys of the field by Anderson and Seacat, in 1957, demonstrates this speed and acceleration with some numerical data.[23] Anderson and Seacat start by looking back a little further than Hall had:

> The application of behavioral science research concepts and techniques in the social and economic aspects of the health field is not new in this country or in Europe, but the momentum with which sociologists, social psychologists, and social anthropologists are being brought into this growing research area is a new phenomenon, and has taken place mainly since 1945.

They then report that the 1956 edition of *An Inventory of Social and Economic Research in Health*, published by their organization, the Health Information Foundation, "lists almost 500 research projects as completed or in progress during that year." Obviously these are not all in the field of medical sociology, however broadly defined, and obviously also, a thorough census was bound to find projects that a more informal survey like Hall's would miss. But still, the increase in rate of work is very large. Finally Anderson and Seacat report on personnel: "Through research inventories and personal contacts and knowledge," they say, "216 behavioral scientists were identified as being engaged in the health field full-time or part-time." It was from these people that Anderson and Seacat collected some questionnaire data we shall discuss later.

By coincidence, another survey of the field was published at about the same time by Robert Straus, who did not know of the Anderson and Seacat survey and who was the agent of what he describes as "a small group of medical sociologists and physicians who met informally in Washington in September 1956, during the meetings of the American Sociological Society," as it was then called.[24] Note, first that this may be the first publication actually to use the term "medical sociology." Note also that Straus reports the beginning of organization for the field, in an "informal Committee on Medical Sociology." Finally, note that the compilation of a directory of medical sociologists, carried out by Straus on the committee's commission, contains 110 "individuals whose basic professional identification is sociology" out of the 162 individuals in the total list drawn up by the committee. In his conclusion, Straus remarks on the rapid rate of change in the field: "From the foregoing summery," he says, "it is apparent that there is a large and varied activity in this field. The field is, however, developing and changing very rapidly, so rapidly that any attempt to describe it runs the risk of early obsolescence."

By 1960, "according to the 'List of Medical Sociologists' compiled annually by the Committee on Medical Sociology, 309 individuals currently defined themselves as engaged to some extent in activities which include medical sociology."[25] Of these, 309,224 were sociologists. In 1960, also, the informal Committee on Medical Sociology was transformed into the formally organized Section on Medical Sociology of the American Sociological Association. In 1962, in a second important survey, Anderson and Seacat sent questionnaires to the 738 members of this section, as of May 1961, about 550 of whom were primarily identified with sociology rather than with social psychology or anthropology or other behavioral sciences.[26]

In this survey, Anderson and Seacat also reported, as an indicator of rapid growth in the field, that of all the courses in medical sociology being given in 1962, only 5 percent were in existence by 1950. Almost half were first offered between then and 1960; and another third had been introduced since 1960. Almost all of these courses were described as being permanently scheduled in the curriculum of some teaching institution. About this time, also, the volume of research and the number of workers in the field were sufficient to justify the foundation of a new quarterly, *The Journal of Health and Human Behavior*. And, finally, the latest survey of medical sociology calls it perhaps the "fastest growing" subfiled of sociology as a whole.[27]

## 2. The Type of Change

On the problem of the type of change this increased rate of work in medical sociology represented for those who came into the field, there is, unfortunately, little direct evidence. As we shall see in discussing new recruits to the field in detail later, however, all the surveys agree that younger, newer men, those who received their degrees since 1945, are the great majority of specialists in this field. We can infer that many of these men have been in medical sociology from the beginnings of their own research and teaching career. Just how many though, we do not know, since many of the younger men started in some other field and then shifted over, probably fairly easily, to medical sociology. For example, in their 1957 survey, Anderson and Seacat report that 28 percent of their respondents said that "a research career in the health field" had not been their goal during their research training. As for the older men, for some it may have involved a basic shift of type of sociological interest; for others a minor shift, say from the field of professions in general into the subfield of the medical profession; and for some few others, of course, who had started out in the field, no shift at all in the type of work.

## 3. The Sources of Change

Now let us look at the several sources of change for medical sociology, remembering, of course, that they have worked not separately, but in combination.

*New ideas and new methods.* New sociological ideas and methods have not been particularly important as sources of the recent great change in medical sociology if by "new" one means ideas and methods developed especially for the peculiar sociological problems of the field. The ideas and methods that have been useful have been "new" only in the novelty of their systematic and often replicated application to medical sociology's problems. Such ideas as "role," in the general sense; "professions"; "social stratification" and "social classes"; "detached concern"; and many others; and such methods as "participant observation," "survey research," and "panel techniques" were all ideas and methods developed in other areas and further applied to medical sociology when other changes occurred that provided opportunities for that application in this field. Anderson and Seacat, in their 1957 survey, for example, report that 62 percent of their respondents checked, as one among several different incentives to go into medical sociology, "the opportunity to apply, test and develop behavioral science knowledge, theory, methodology and hypotheses."[28]

*New access.* A change in the attitude of medical personnel, especially in the attitude of certain key physicians and administrators, toward behavioral science research seems to have been one of the factors chiefly responsible for the great growth of medical sociology. This change of attitude, often manifested in actual research collaboration with social scientists, has resulted in the new access to hospitals, to patients, to their families, to medical students, and to the doctors themselves without which medical-sociological research would have been impossible. What brought about this change, how far it extends in the medical world, and how deeply, however, are matters on which we have no systematic evidence. One impression I have is that those physicians who saw the patient as "a whole person," not just an organic case, and those who were most science- and research-oriented were most likely to look to the behavioral sciences for expert aid.

*New recruits.* As we have already seen, new recruits to the field of medical sociology have definitely been among the agents of the changes it has undergone. Most of these recruits have been younger men, but some older men have also switched over, though how many of these there are we do not know. In their 1957 survey, Anderson and Seacat reported that 58 percent of their respondents had received their Ph.D.s in the previous six years. We are also told that 53 percent of the researchers had been in the health field two years or less, but we are not told in what measure this group overlapped with the group of recent Ph.D.s.[29] By 1962, in their second survey, Anderson and Seacat reported that the proportion taking their Ph.D. in 1950 or later had risen to 75 percent. In this new field, at this time, almost half of the respondents were still young men, in the thirties. "The newness of the field and the youth of the participants in it," however, as Freeman, Levine, and Reeder remark, "is balanced by the active engagement in socio-medical problems of many of the leading and relatively more elder statesmen in sociology. For example, Merton's studies of medical education, Parsons's analysis of the role of the patient, and Hughes's observations on the medical professions . . ."[30] represent the work of distinguished older sociologists. But all three of these men had been interested either in the the professions in general or in the medical profession specifically (for Parsons both) before the recent enormous rate of increase of medical sociology. Their interest in the field represented a less basic type of change than it may have for some other older men.

*New funds.* Clearly, new funds were important in the changes occurring in medical sociology. We know, for example, from the first Anderson and Seacat survey, that 15 percent of the respondents were willing explicitly to say that "the availability of research funds" had been an important incentive for their entrance into the field. Also, we know that new grants to the medical schools and to the universities by various foundations, perhaps especially the Russell Sage and the Commonwealth foundations, were indispensable to the growth of medical sociology.[31] Indispensable also were new funds from the federal government. Within the federal government, as Williams has suggested, "the National Institute of Mental Health plays the major role."[32] State, county, and city governments have also played some part in supplying new funds.[33] However, as Williams further remarks, we have no "hard data" either on the overall amounts and increases of funds for medical sociology, nor for the varying proportions of those funds coming from these several different sources, the foundations and the various levels of government.

*New professorships and other positions.* Many of the new funds have been used to establish new professorships and new research positions for medical sociology, and these have served to attract new men to the field. Some 24 percent of the respondents in the 1957 Anderson and Seacat survey checked "professional opportunities" such as new positions among their reasons for entering medical sociology. In the Straus survey, of the 110 respondents whose basic professional identification was with sociology and for whom data were collected, 34 had their primary research or teaching position with an academic department of sociology, and all the rest were with medical schools, research organizations, or foundations, most of these being new positions. In the medical world, medical sociologists "are regular members of departmental units in public health practice, preventive medicine, epidemiology, biostatistics, and psychiatry," Freeman and his colleagues tell us, and nearly all of these are attractive new positions.[34]

*Social problems.* Lastly, "social problems" in relation to the medical world have been among the sources of growth for medical sociology. Among the researchers themselves, according to the Anderson and Seacat 1957 survey, some 39 percent said "the opportunity to deal with problems of vital importance to human welfare" was among their incentives. Health in general and mental health more particularly in recent years have come to be defined as important social problems in a society like ours. This new or more sharpened definition has been important in the expansion of all medical resources and facilities and also in that of medical sociology as a part of those resources and facilities.

## 4. Functions and Dysfunctions of the "Fashion" for Medical Sociology

Without further research it is difficult to say anything very precise about the functions and dysfunctions of the recent changes in medical sociology. Again, on impressionistic grounds, it would seem that a good deal of the research and teaching in this field has been both of practical, applied usefulness to various types of personnel in the medical world and of more fundamental scientific usefulness for the accumulation of empirical generalizations and the sharpening of theoretical and methodological tools in sociology itself. But these are matters that require closer scrutiny and study than they have yet been given. On the side of dysfunctions, it is difficult to see anything of consequence, though there may be some sociologists and some laypeople who think it would have been better to turn the efforts that have gone into medical sociology, or some of them, into more important theoretical or practical problems, as they define "importance."

In conclusion, I will hardly need to note again that these have merely been "notes" on the "fashionableness" of medical sociology during the last twenty years. It is not hard to be convinced that further research is necessary if we are to understand both this particular "fashion" or change in general. I hope that such research will be forthcoming before long, and also research on many other instances of "fashion" or change in the social and natural sciences.

## Notes

1. Bernard Barber and Lyle S. Lobel, "Fashion in Women's Clothes and the American Social System," *Social Forces* 31 (1952): 124.
2. But see a useful beginning of such analysis in Warren O. Hagstrom, *The Scientific Community* (New York: Basic Books, 1965), pp. 177–84.
3. Herbert Menzel et al., *The Flow of Information among Scientists*, unpublished report, Bureau of Applied Social Research, Columbia University, May 1958.
4. Ibid., p. 42.
5. *Science* 141 (1963): 765.
6. Unsigned editorial, *Science* 139 (1963): 377. In science, of course, "glamour" is no virtue.
7. On the exponential growth rates of science, see D. J. de Solla Price, *Little Science, Big Science* (New York: Columbia University Press, 1963). On the inevitability of revolutions in science, see T. S. Kuhn. *The Structure of Scientific Revolutions* (Chicago, Ill.: University of Chicago Press, 1962).
8. Kuhn, *Structure of Scientific Revolutions*.
9. Gerald Holton, "Scientific Research and Scholarship: Notes toward the Design of Proper Scales", *Daedalus* 91 (1962): 385.
10. Ibid.
11. Ibid., p. 390.
12. See Bales, *Interaction Process Analysis* (Cambridge, Mass.: Addison-Wesley Press, 1952); see also, P. Hare, E. F. Borgatta, and R. F. Bales, eds., *Small Groups* (New York: A. A. Knopf, 1955), which includes an extensive bibliography describing the new "fashion" as well as its antecedents.
13. R. H. Shryock. *The Development of Modern Medicine* (New York, 1947).
14. Holton, "Scientific Research and Scholarship."
15. E. M. Rogers, *Diffusion of Innovations* (New York: Free Press of Glencoe, 1962), pp. 36–37.
16. Editorial matter in *Science* 142 (1963): 1153.
17. Joseph Ben-David and Awraham Zloczower, "Universities and Academic Systems in Modern Societies," *European Journal of Sociology* 3 (1962): 54–56.
18. Mayr, *Science* 141 (1963): 765.
19. On the concept of sociological ambivalence in general, see R. K. Merton and E. G. Barber, "Sociological Ambivalence," in *Sociological Theory, Values, and Sociocultural Change: Essays in Honor of Pitirim A. Sorokin* (New York: Free Press of Glencoe, 1963). See also R. K. Merton, "The Ambivalence of Scientists," *Bulletin of the Johns Hopkins Hospital* 112 (1963): 77–97. Among the nine different patterns of ambivalence Merton discriminates, there is the following, (p. 78): "2. The scientist should not allow himself to be victimized by intellectual

fads, those modish ideas that rise for a time and are doomed to disappear. BUT he must remain flexible, receptive to the promising new idea and avoid becoming ossified under the guise of responsibly maintaining intellectual traditions."

20. T. Dobzhansky, "Evolutionary and Population Genetics," *Science* 142 (1962): 1131.

21. Ernst Mayr, editorial in *Science* 141. For other expressions of ambivalence about "fashion," see the letter by E. D. Hanson about Mayr's editorial in *Science* 141 (1963): 623, and Honor B. Fell, "Fashion in Cell Biology: The Motives That Prompt Us to Follow Fashions in Research Are Various and Not Always Estimable," *Science* 132 (1960): 1623–27.

22. Oswald Hall, "Sociological Research in the Field of Medicine: Progress and Prospects," *American Sociological Review* 16 (1951): 639–44.

23. O. W. Anderson and Milvoy Seacat, *The Behavioral Scientists and Research in the Health Field: A Questionnaire Survey*, Health Information Foundation, Research Series 1 (1957).

24. Robert Straus, "The Nature and Status of Medical Sociology," *American Sociological Review* 22 (1957): 200–204.

25. Samuel W. Bloom, Albert F. Wessen, Robert Straus, George C. Reader, M.D., and Jerome K. Myers, "The Sociologist as Medical Educator: A Discussion," *American Sociological Review* 25 (1960): 95–101.

26. O. W. Anderson and M. S. Seacat, *An Analysis of Personnel in Medical Sociology*, Health Information Foundation, Research Series 21 (1962).

27. See H. E. Freeman, S. Levine, and Leo G. Reeder, "Present Status of Medical Sociology," in an anthology edited by these three men, *Handbook of Medical Sociology* (Englewood Cliffs, N.J.: Prentice-Hall, 1963).

28. Anderson and Seacat, *The Behavioral Scientists and Research.*

29. Ibid.

30. Freeman, Levine, Reeder, "Present Status of Medical Sociology."

31. See Straus, "Nature and Status of Medical Sociology."

32. Richard H. Williams, "The Strategy of Socio-Medical Sociology," in Freeman et al., eds., *Handbook of Medical Sociology*, pp. 441ff.

33. Ibid.

34. Freeman, Levine, Reeder, "Present Status of Medical Sociology."

# 7

# Trust in Science

## Trust: Definitions and Propositions

Trust is a generic aspect of all social interaction and all social systems. Even though it is, therefore, a basic social phenomenon, it has very frequently been used ambiguously by past social thinkers, by the man-in-the-street, by journalists, and by contemporary social scientists.[1] The ambiguity and the vagueness about the meaning of trust are manifested in the widespread use of many apparent synonyms and antonyms, often themselves undefined: honesty, faith, confidence, alienation, anomie, malaise. The result is a conceptual morass. To put us on more solid analytical and empirical ground, we need to examine trust in the light of our general understanding of social relationships and social systems. The construction resulting from this examination should, of course, be empirically usable and testable. Very briefly, I offer just such a construction.

Social interaction consists of an endless process of learned, negotiated, confirmed, and disappointed expectations that individuals and other social actors, such as groups and organizations, have of one another. Expectations are the basic processual stuff that gets structured into social roles and social systems. These structures are never finally fixed but are only temporary ways of seeing some relatively constant patterns in the expectations that actors have of one another. Expectations can be cognitive, emotional, moral, or some mixture of these responses.

In terms of the expectations that make up all social interaction, trust has at least two essential meanings. One meaning is expectations of technically competent role performance. We trust, in this sense, that a role incumbent knows and can execute the technical standards prevailing in his or her role. This meaning of trust is important in all societies, but is perhaps especially so in a society like ours where there is such a vast accumulation of knowledge and technical expertise based on that knowledge. Scientists very much expect that a certified scientist can be trusted in this sense: to know the technical standards relevant to his or her work, to be a competent performer.

A second meaning of trust is expectations of fiduciary obligation and responsibility. We trust that the role incumbent will fulfill his or her duty in certain situations to place the values and interests of his or her role partners, or the collectivity of which they are all members, above the incumbent's own immediate interest. As trusted persons in this sense, observing their fiduciary obligations, it is to their moral interest to put their other interests second. Again, this kind of trust is important in all societies but may even increase in importance in a society like ours where greater complexity, differentiation, and greater possibilities of individual power enlarge the need for responsibility for others and for the collectivity as a whole. Scientists very much expect that a certified scientist can be trusted in this sense too, to observe the moral norms of science, to fulfill one's normative obligations to co-workers, to the scientific community, and to the public as well. Trustfulness, trustworthiness, trust in both senses are indispensable to effective science.

These two analytic types of trust are independently variable; an actor and a group may measure not only the same on both but also higher and lower on one or the other. This is certainly the case in science. To say of an individual scientist, for example, that "he is too clever by half" is perhaps to say that we trust his technical competence more than we trust his fiduciary responsibility.

Moreover, we have to note that the relative emphasis on each of the two types may vary from one social institution to another and from one time to another. Thus market institutions in our society, although relying a great deal on both kinds of trust, still place more reliance on price mechanisms, on the market, to control behavior than do professional institutions such as science and medicine. The professional occupations rely heavily on trust of both kinds to ensure effective performance and to forestall deviance.

Even in the same institution, the emphasis on the importance of the two types of trust may vary over time. As recently as the late 1960s, medical scientists did not give as much salience as they now do to fiduciary responsibility for the human subjects of their experiments. There was not high concern for getting informed consent from these subjects and for making sure that the cost/benefit ratio of the research was favorable for the subjects. In a study carried out in the late 1960s, my colleagues and I asked a sample of some 350 medical scientists, all of whom used human subjects, this question: "What three characteristics do you most want to know about another researcher before entering into a collaborative relationship with him?" The responses showed a preponderant emphasis at that time on technical competence and technical trustworthiness. Eighty-six percent of the respondents mentioned "scientific ability," 45 percent "motivation to work hard," and 32 percent "intellectual honesty." As for fiduciary trustworthiness toward the subjects of research, only 6 percent mentioned "ethical concern for research subjects."[2]

Although, regrettably, we have no later studies to bring ours up to date, it is very likely that nowadays the responses would give more weight to fiduciary trustworthiness toward subjects. This is so because of the scandals that have revealed the unethical use of human subjects, scandals that have brought this problem of medical-science untrustworthiness to wide professional, governmental, and public attention. Medical education now pays somewhat more attention to moral trustworthiness toward subject and patients than it used to. Finally, and perhaps most important, all scientists using human subjects must now have their research protocols approved by local ethical screening boards, generically known as IRBs, or institutional review boards. These IRBs were mandated by the National Institutes of Health in 1986, on pain of loss of funding for noncompliance.[3]

Another example of change over time within science is the greater salience that fiduciary obligation to one's immediate colleagues and to the scientific community as a whole has come to have in recent years because of the widely publicized cases of fraud in science that have occurred at major research universities such as Harvard, Yale, Columbia, and at the Sloan-Kettering Institute. We shall look at these cases in detail later.

Both kinds of trust have important functions for social interaction and social systems. They contribute to orderliness, effectiveness, and social control. But they have their dysfunctional consequences as well. When trust is abused, then we may have deviance, corruption, and conflict. Hence the need for *functional complements* and *functional alternatives* to trust. Trust alone may not be, sometimes is not, enough for effective social action and control. These necessary complements and alternatives may be either informal social processes such as ridicule, unhelpfulness, or ostracism, or they may be formal entities and processes such as credentialing committees, ethics committees, monitoring, insurance, or legislation and legal processes.

Within science itself, there has been a strong tendency to rely on trust itself and on informal complements and alternatives. However, as untrustworthiness has seemed to increase and has become public knowledge, scientists, funding agencies, and the attentive public have proposed the use of more formal complements and alternatives. In science, as in other social systems where the ideal that Talcott Parsons called "the company of equals pattern" prevails, universal and prevailing trust is preferred. Informal control mechanisms are used but not admired. Formal social controls are abhorred, seen as an affront to the valued solidarity and cooperation among equals. Trust has an expressive as well as a control function. Stable trust expresses and maintains the shared values of a social system such as science. When trust breaks down, when it seems that the shared values of science no longer exist, then scientists experience poor morale, mutual suspiciousness, overt hostility, and conflict. All of this makes impossible the easy cooperation and trustworthiness in both senses that is essential to effective performance as scientists. Excessive competition among scientists and among their laboratories may result in distrust and its negative consequences. Such seems to be the case presently for the American and French laboratories that are competing in the field of AIDS research.

The scope and limits of both trusted competence and fiduciary trustworthiness are often hard to define, changing as the situation and the partners in the trusting interaction change. There is some tendency, therefore, for trusted actors to feel they are not being given enough trust, to overreach as to their claims to either competence or ostensible fiduciary performance. Some scientists have assumed attitudes of omni-competence, as when Nobel Prize laureates, encouraged, to be sure, by the media and the public, make pronouncements on scientific and social problems and policy on which they have no special knowledge or competence. On occasion, some scientists have felt that they were beyond moral questioning or reproach. For example, American molecular biologists, themselves somewhat fearful at one point of the possible harmful consequences of gene-splicing, gathered together at Asilomar in California to write a set of protective rules for work in their field. This was a high form of scientific trustworthiness. But when citizens at both the national and the local levels (as in places like Ann Arbor, Michigan, and Cambridge, Massachusetts) protested this "usurpation" of their rights in a participatory democracy to be included in decision-making processes about a matter so important to the public welfare, some of the leading molecular biology scientists thought this an attack on their fiduciary trustworthiness. They expressed their regret at having informed the public at all. Since no one group of specialists has a monopoly on morality and fiduciary trustworthiness, the scientists should have welcomed responsible public participation.

Since omni-competence and fiduciary perfection are rare in individuals and groups, since there is always some lack of trustworthiness in either respect, there is a constant need in social interaction in all spheres for what I have called "rational distrust." Fellow actors or fellow participants in social systems may well be correct, sometimes, in their judgments that distrust is the appropriate and rational response to certain claims of competence or fiduciary responsibility. Not every expression of distrust against some scientist or scientific claim is, as some scientists have held, an indication of public ignorance or antiscientific values. All specialists in a complex society, and scientists are specialists par excellence, must expect occasionally to be questioned or held to account with regard to both competence and fiduciary concern for the public interest and welfare. Of course, where public distrust is based on ignorance or wrong values, then scientists have the right and duty to combat it with full force.

One last general point about trust. Like all social mechanisms and structures, trust is never fully and finally established. Trustworthiness in science, of both kinds, must be displayed and proved continuously. Maintaining trust is an endless, if sometimes tiresome task. Within science itself, with attentive publics, and with the public at large, scientists have to keep working at maintaining not only the appearance but the fact of trustworthiness.

### A Brief Historical Note

Trust within science, and then eventually public trust in science, depends on the establishment and maintenance of certain technical and moral norms as standards for behavior. The *moral* norms—such as organized skepticism, communism (communalism), and disinterestedness—were first explicitly and systematically discussed by Robert Merton.[4] Despite criticism from those who deny the existence of normative elements in any social interaction, those who think that alleged norms are merely "rationalizing ideologies" used to justify "interests," and despite various useful qualifications to the Mertonian formulation, Merton's statement of the moral norms remains essential for the sociology of science. The *technical* norms—such as the standard of logical reason, and testing by empirical evidence preferably through the use of controlled experiment—have long been the focus of discussion in the philosophy of science, now a highly specialized branch of philosophy in general.[5] Both of these types of norms create the necessary conditions for the competent technical performance and the fiduciary responsibility that are essential for science.

According to an interesting historical account by Joseph Ben-David,[6] these two kinds of norms first emerged from embryonic prototypes into the requisites of an explicit role in the seventeenth century. Scientific roles in that century were primarily recognized and supported in an informal network of "scientists," a term that was actually not coined until the 1830s by William Whewell. That network was described by its members as "an invisible college." Although informal networks and interaction are preferred in the actual work situations of scientists, it was soon recognized that some formal association might be helpful to gain public recognition, legitimacy, and financial support. Hence the establishment in Great Britain of the Royal Society. The Royal Society, and many subsequently established scientific academies and professional associations, usually with their own refereed publications, became the functional complements to informal interaction among scientists for maintaining trust in both the competence and the fiduciary responsibility of scientists.

Ben-David stresses the importance of the strengthening of the old academies and the emergence of new ones at the end of the eighteenth and into the nineteenth centuries. He attaches so much importance to this indispensable development for the maintenance of trust in science that he calls it "the Eighteenth-Century Revolution in Science." Napoleonic initiatives and reforms were a key element in this "revolution." During the nineteenth and twentieth centuries, scientific academies transcended more and more their national boundaries and widened the scope for scientific trust. On the institutional side, there has been no further "revolution" to compare with that of the eighteenth century. The established and strengthened informal and formal social mechanisms of science have provided the conditions of trust that have made all the more substantive (cognitive) discoveries and revolutions possible.[7]

## Trust within Science

Trust within science depends on the continuing and effective fulfillment of both the technical (cognitive) and the moral norms that structure scientific behavior. Fulfillment of the technical norms is taken as satisfactory technical competence and establishes trustworthiness in that respect. And fulfillment of the moral norms shows satisfactory fiduciary responsibility to the community of scientists and earns a reputation for trustworthiness in that respect. *Per contra*, failure or violations in either respect are defined as untrustworthiness and lead to informal sanctions from one's peers. In what is considered a "crisis" situation, however, as we shall see, there may be a call for more formal control mechanisms and sanctions, such as loss of job or withdrawal of research funds by the funding agency.

Violations of both kinds occur in science. I first consider delinquencies or shortcomings in regard to technical norms, then in regard to moral norms. Since Harriet Zuckerman[8] has published an excellent review and analysis of both kinds of scientific untrustworthiness or deviance, I shall rely on her account while extending its analysis and bringing it up-to-date with more recent materials.

The scientific errors that occur as a result of technical incompetence are roughly classified by Zuckerman into "reputable" and "disreputable" errors. The former are those that occur sometimes and to all scientists despite their having maintained high standards with regard to the general technical norms that are relevant to all science and to the somewhat more specific technical craft procedures that prevail in the endless specialties into which scientific work is divided. Errors of the latter type result from gross neglect or violation of either general or specific technical norms. Reputable errors are accepted as part of the game, the inevitable slips that happen to even the star players. They may occasion informal control mechanisms, such as jokes or teasing, but not charges of real untrustworthiness. Disreputable errors arouse much stronger feelings and moral concern among scientific peers. Scientists will express contempt and anger at such inexcusable incompetence. Disreputable errors waste time and resources of those who are trying to replicate studies or to build new research and analysis on them. They put untrustworthy data into the data bank on which scientists in a specialty depend. Such data depositors are to be despised if their errors cause a lot of trouble and they are repeated. Probably errors are all the more blamed when they occur in connection with possibly important findings. In a great deal of science, experiments are not repeated by others; they are taken on trust. This is especially true for routine and minor work. But where work seems important, scientific peers pay more attention and are more likely to try to replicate or use the findings. Contrariwise, errors connected with minor "discoveries" may never be found out because of the lesser attention paid to these findings.[9]

While the indignation that is provoked by the untrustworthy incompetence of disreptuable errors indicates that even the technical norms of science are supported by moral norms, even stronger moral condemnation comes down on those scientists who are untrustworthy with regard to the central moral norms of science such as organized skepticism, universalism, communalism, and disinterestedness. Resort by scientists to *ad hominem* attacks on their peers, plagiarism, secrecy, and fraud are types of moral delinquency that violate one or more of these central moral norms of science.

Plagiarism, for example, is a violation of the norm of communalism. That norm prescribes that all scientists must contribute willingly to the common store of knowledge and anyone who is able to do so may draw from that common store. The rule may be phrased "from each according to his production, to each according to his needs for the knowledge that has been produced." There is no property in scientific knowledge except the "intellectual property" that may be thought to exist in the credit and prestige that peers give to the producers of knowledge in recognition of their achievements. Plagiarism violates the norm of communalism because the plagiarist is falsely appropriating credit that should have been given the rightful producer of valuable knowledge for the common store. Plagiarism is often imputed to opponents in the priority quarrels that dot the whole history of science and that Robert Merton has analyzed so brilliantly.[10]

Secrecy, similarly, is a violation of the communalism norm. The secretive scientist keeps from the common pool of knowledge findings and techniques that are necessary for further enlargement of that common pool. Individual scientists or laboratories that know themselves to be in competition may feel under some pressure to be secretive rather than open and cooperative.[11] It is, of course, often difficult to specify just when secrecy in science is occurring. What looks to outsiders like secretive and untrustworthy withholding by a scientist may look to the scientist like proper caution and concern about the validity of one's results; the scientist will not disclose until he or she is sure of them, and perhaps also, sure of getting proper credit. Secrecy, then, is more a relative than an absolute matter; that is why scientists speak of "undue secrecy."

Fraud, the calculated deception of one's peers in science, is a violation of the norm of disinterestedness and results in the strongest condemnation for fiduciary untrustworthiness. Scientists observing the norm of disinterestedness put the welfare of the collectivity of scientists above their own self-interest. Obviously the fraudulent scientist puts self-interest in gaining recognition at any cost first.

Fraud in science takes a variety of forms. There may be outright fabrication of data; there may be manipulation of data to make them come out the way the scientists want them to; there may be suppression of data that would falsify or call into question the scientist's findings, and there may be misappropriation of data and of credit for work done. Something like all of these fraudulent behaviors were discussed about 150 years ago by Charles Babbage as "forging," "trimming," and "cooking."[12] So fraud has been around in science a long time. But what its present structural sources, its magnitude, and its incidence and distribution are is not entirely clear, although we now have some intensive case studies and a few data on fraud in science as well as many assertions not well supported by good research.

A special session at the 1985 Annual Meeting of the American Association for the Advancement of Science (A.A.A.S.) viewed with alarm what it felt was a great increase of fraud in science, an increase that was said to be destroying the moral integrity of science and to be eroding "the moral fiber of our future physicians," that is, the ones engaged in biomedical research.[13] Most of the participants were from the biomedical sciences, and that is not surprising, since it is in those fields that there is now a great deal of basic science activity and competition; they are the current "hot" fields in science.

The participants seemed all to be in agreement that competition in their fields is now the main structural source of fraud. Robert G. Petersdorf, dean of the School of Medicine at the University of California, San Diego, who has long been a scientific and moral leader in American medicine, said: "It is true that science in 1985 is too competitive, too big, too entrepreneurial, and bent too much on winning." And Hendrik Bendixen, dean of the College of Physicians and Surgeons, Columbia University, in whose medical school two medical researchers were recently required by an investigating committee to retract nine published papers because of apparent manipulation of data, concurred by affirming that "the high pressure environment" at the medical schools of major universities is responsible for the current cases of fraud. There is too much competition for scarce research funds and to achieve the required volume of publications, often just by volume it is alleged, for promotions and tenure.

Petersdorf and his fellow participants feel that this competition is especially harsh within and between the large laboratories that are now so important in the biomedical as well as other sciences. Patricia Woolf, a sociologist of science at Princeton University, reported to this A.A.A.S. session that

> fraud has been detected at our best universities, where research excellence is emphasized and where many professors do publish considerably more papers than the norm. . . . young people on the tenure track have been more frequently caught at fraud than older, established scholars. . . . many of the perpetrators of publicly disclosed scientific fraud had published many papers, especially in the time period immediately surrounding the fraud.[14]

Indeed, perhaps the most notorious of recent cases, fraud through manipulation of data in genetic experiments by Dr. William Summerlin ("the case of the painted mice") occurred at a laboratory in Memorial Sloan-Kettering Hospital in New York, a world center in cancer research.[15] In this laboratory, and in others like it at such prestigious universities as Columbia, Harvard, and Yale where cases of fraud have recently been discovered, there is enormous competition for funds and promotion. There is also a bigness that makes it difficult if not impossible for the directors of those laboratories to supervise, monitor, and check the work of their associates properly. The director of Summerlin's laboratory at Sloan-Kettering had put his name on 341 papers in the five years preceding the discovery of Summerlin's fraud. The director of the laboratory at Harvard, where a researcher had falsified data for a research in cardiology, had put his name on 171 papers in the previous five years. And the director at Yale, in the same time period, had presumably authored or co-authored 201 papers. As I indicated in my general remarks on trust, there is always some need for functional alternatives and complements to trust. In these laboratories, these alternatives and complements had broken down. Effective monitoring, auditing, and checking by the director were weak and ineffective. There was too much trust; it turned out to be undeserved, and in at least two of the laboratories not only the untrustworthy scientists but the laboratory directors paid heavily for it in loss of jobs and prestige.

But the problem of violations of trust in large laboratories goes beyond the relations between the director and the senior scientists. The greatly increased division of labor and specialization in "big science" now means that not only must the director check the senior scientists but they, in turn, must check junior scientists and even laboratory technicians. In the Columbia case referred to above, where the inside and outside investigating committees found at least "disreputable error" and perhaps overstrong commitment to favorable findings, it has recently come to light that a major and confessed fault lay with one of the laboratory technicians. Under threat of suit from one of the "guilty" senior scientists, a laboratory technician at Columbia has written a letter acknowledging that "actions I took resulted in the biasing of the experimental data."[16] Laboratories of any size now consist of complex social hierarchies and complex sets of specialists. All this changed structure of science requires more fulfillment of trust than in the past, when simpler structures and a less esoteric division of labor prevailed. Large laboratories must now be stricter in selecting and supervising their members at all levels and in all branches for their trustworthiness.

Despite the great alarm about scientific fraud expressed at the A.A.A.S. meeting, there were no firm estimates of the actual magnitude of current fraud. There is, of course, some tendency among those outraged by even a few cases of untrustworthiness to exaggerate its magnitude. And the larger amount of publicity that has been given to the exposed cases of recent fraud may also magnify the problem among scientists themselves and their attentive public. Speaking for the National Institutes of Health (N.I.H.), which funds most of today's biomedical research, William F. Raub, deputy director, reported to the A.A.A.S. session that cases of fraud coming to the attention of his agency were relatively uncommon, about two cases a month of allegations of possible misconduct. This number, he said, is "a vanishingly small fraction" of the approximately 40,000 scientists who are supported by N.I.H. at any one time. Of course, institutions receiving research funds from N.I.H. are not now required to report instances of fraud, but N.I.H. is about to propose a regulation requiring such reporting. Still, Dr. Raub concluded that he had no reason to think that a large number of fraud cases were presently not reported. It is in the nature of the case, of course, given the predominantly informal control mechanisms over untrustworthiness in science, that it should be hard to make precise measures of the amount of scientific fraud.

Whether there is actually a "crisis" in trustworthiness in science at the present time because of a "large" increase in the amount of fraud, or whether it is only perceived to be one, to some degree makes no difference. Actors in general, and in this case scientists, act upon their perceptions. In the present "crisis" situation, we can expect both an intensification of informal monitoring mechanisms in science, especially at prestige universities and their large laboratories, and the creation of ad-hoc formal mechanisms, such as the N.I.H.-required-reporting regulation and the setting up of special monitoring and investigating committees. So far, no general committee to investigate fraud in science has been established by such a body as the National Academy of Sciences, but university committees to investigate and adjudicate specific allegations of fraud at their own universities have carried out their charges. Unless the situation becomes clearly much worse, this is probably as far as complementary mechanisms for maintaining technical competence and fiduciary responsibility will go.

A great deal of recent discussion of violations of trust within science has concentrated on what happens in the laboratories, in the actual working situations of scientists. But the problem of trust and distrust in science reaches beyond the immediate working situation. When new work is proposed to funding agencies, when articles reporting findings and discoveries are submitted to scientific journals, and when books are offered to publishers, all this proposed and finished work must be evaluated by "judges," "referees," "reviewers," acting singly or in panels and who are usually the "peers" of the scientist or laboratory director whose work is being evaluated. A great deal obviously depends on the trustworthiness of these peer evaluators of scientific contributions; they must be both competent and fiduciarily responsible to the welfare of science. While no "crisis" of trust is currently perceived in this area of science, and while probably, in the aggregate, trustworthiness in both senses prevails, still if one listens to the "murmuring of the masses" within science, there are endless complaints about the lack of competence and the self-interestedness on the part of evaluators of all kinds. Almost every working scientist, in both natural and social science, has a tale to tell about how some work of his or hers, some work of an associate, has been incompetently reviewed or has been mistreated by some reviewer with a self-interested motive. Not all of these complaints are valid, of course, but they indicate the presence of some continuous amount of distrust and its consequent sense of grievance and hostility in science. This distrust is clearly manifest in the controversy that frequently recurs over the principle of anonymity in the evaluation process. Some scientists hold that anonymity increases the possibility for incompetence and fiduciary irresponsibility to occur; others, in direct disagreement, say that only anonymity makes it possible for the evaluator to be candid in judging the quality of the work and to think first and only of the good of science. Given this disagreement, it is unlikely that all distrust of the peer-reviewing process, which is so essential in science, can be eliminated, but it is clear that the efficiency and trustworthiness of the process is a matter that is of great concern and will recurrently itself be evaluated.

## Public Trust toward Science

The problem of public trust toward science derives from the enormous power that science now has in our society. It is not the power that political and economic actors have, though scientists now are often trusted advisers to such actors and may even themselves serve in these positions. It is, rather, the power of knowledge, the power not only to create constant technological innovation and its vast consequences for good and ill but also, by ever new knowledge of the physical, biological, and social worlds, to change the frames of our basic meanings in every aspect of culture.[17] In the postwar world, science has grown exponentially in both numbers and power.[18] Science has discovered new sources of energy, taken us into new worlds of "space" and the universe more generally, created "miracle drugs," and taught us new ways of seeing and judging our behavior. Through all the multifarious benefits it has brought, science has increased public trust certainly in its competence and also, but more ambivalently, in its fiduciary responsibility for the public welfare.

Great power is seldom absolutely trusted, and so it has been with science. There are three general reasons why science, and more generally the practicing professions that depend on science for their knowledge base, have recently become objects of public distrust, some of it rational, some irrational, three reasons why science has become partly a "social problem" as well as a social boon. The first reason, of course, is just that very great power that science now possesses to do good and harm. It is a power that gets credit for "the green revolution" in agriculture but that also gets blamed for the harmful consequences of the discovery of atomic energy and the possible and imagined consequences of human gene-splicing. The public is uneasy in the presence of very great power of any kind, especially in times of increasing power, as is the case now for science. It is a time before the public can test how much science can actually be trusted and before it can devise those functional alternatives and complements to trust toward science that are required for effective social control.

Some empirical evidence on the public's perceptions of the power of science and its consequent feelings of trust and distrust can be gleaned from the data that have been collected by American survey researchers during the last twenty years. Fortunately, these data have been carefully collected and analyzed by Lipset and Schneider in their book, *The Confidence Gap*.[19] These "confidence in institutions" studies reviewed by Lipset and Schneider, unfortunately, do not use questions that distinguish clearly the two types of trust I have defined here, trust as technically competent performance and trust as fiduciary responsibility. This is necessary because of the independent variability of these two types. Nor do the studies distinguish between the institutional systems themselves and the current incumbents of positions in those institutions. It would be good if, in future studies, such distinctions were made by the questions asked; it would be easy to do so. It would give us a better picture of the public's trust and distrust in science and other institutions. Still, despite these limitations, some useful points may be inferred from the data as they stand.

The massive data from the surveys shows that there are mixed feelings of trust and distrust toward all the powerful institutions in American society: government, labor, business, science, medicine, the military, and religion. There is some lack of "confidence" in all of these institutions. As Lipset and Schneider put it, in agreement with what I have said above about the general distrust of all great power, "the simple fact that elites have power makes them a source of suspicion and distrust to the American people."[20] But some institutions earn more "confidence" than others. Business, government, and labor earn less than science, education, and medicine. We may surmise that there are two reasons for this. First, the great power of business, government, and labor is more visible to people; they have learned to be more aware of it than they are of the power of science, although surely nowadays no one could be completely unaware of that power. And second, science, medicine, and education are perceived to be more trustworthy in terms of fiduciary responsibility to the public welfare. Again, not that they are seen as absolutely trustworthy in this regard, but that they are more so than business, government, or labor. Both kinds of trust are never absolutes, only relative. Despite its ambivalence toward science, the public feels it can trust it relatively more than some other institutions. Trust in this case, and in general too, is always a relative matter.

A second and interrelated reason why the trustworthiness of science is now a social problem is the greatly increased emphasis we have recently come to put upon the fuller realization of one of our old and central values, the value of equality. The public wants more of the good things of life, not just material goods, as is so frequently alleged, but all the social goods, the health and happiness, the leisure and self-fulfillment that our society seems able to provide. People want all these goods distributed a little more equally than they have been in the past, and so they want the various forms of power that produce and control these goods distributed a little more equally too. They want somewhat more control over not only politicians and businessmen, therefore, but also over scientists. Whether it is the young against the old, blacks against whites, students against teachers, patients against doctors, or the public against scientists, all those who feel themselves unequal either as to opportunity or as to outcomes in resources, goods, and power express some distrust of both the competence and the responsibility of the high and mighty—and all this in the name of the value of equality. This is a value, we must remember, that vibrates in our society for the more powerful and well-off as well as the less so. Powerful scientists seek the public trust not only to gain more funds for their work but in terms of such values as equality. They want legitimacy as well as resources.

Finally, because of its own increased education and resulting competence, the public engages in ever more testing of the actual trustworthiness of scientists and other professionals. Much to the horror of many older scientific and other experts, the public is now less passively deferential to them. When scientists make excessive claims to trustworthiness, the public wants to do more monitoring of their claims and action. In public controversies over atomic energy plants or over the university or industry locations of gene-splicing, the public brings in its own scientific experts to monitor or counter the views and actions of the established experts.[21] Of course, by itself, the public has no technical competence in science, but it can exercise some control by such complementary mechanisms as required public discussion among differing scientists. Even this has its limits, because the public cannot decide among differing scientists. Ultimately, there must be trust toward science and its internal control mechanisms.

In addition to these general sources of public distrust of science, there are some more specific ones. Public distrust arises from the conviction that too often the scientists' means become ends, that their concern for technically competent performance overrides the need for the fiduciary responsibility to the public welfare that this performance is supposed to serve. In Robert Oppenheimer's classic description of this turning of means into ends in the physical sciences, he said that the application of discoveries in atomic science to the making of atomic bombs was "too sweet" a problem in technical terms to be given up in consideration of the dangerous social consequences that success in such work might bring. Similarly pointing to this phenomenon of means into ends, at the time of the dispute over the first research on gene-splicing, Erwin Chargaff, the Columbia University biochemist, himself a distinguished contributor to molecular biology, denounced what he called "the devil's doctrine" that what *can* be done in science *must* be done. Biomedical scientists who formerly did dangerous experiments on human subjects without their informed consent were also turning their means into ends, without regard to the ends important to the public who furnished the subjects.[22]

Another specific source of public distrust of science comes from the frequent claim that science makes for its absolute autonomy. Such claims are not unknown from all kinds of specialists. I have already mentioned that some sections of the public came to view as arrogant the claims of the DNA scientists meeting at Asilomar to an absolute autonomy for their views on the dangers and regulation of their new science. Criticism of the arrogance of scientists is not new in our democratic society. In his distinguished sociological history of American physics, Daniel Kevles of the California Institute of Technology has documented in detail the "political elitism," as he calls this arrogance, of organized American scientists from the Civil War to the present.[23] The scientists have often wanted, and have recently achieved, he feels, an autonomy in American society probably unmatched by any other powerful social group. Concluding his judicious appraisal of this source of public distrust of science, Kevels asks: "How was the scientific community's demand for political elitism to be reconciled with the principle of politically responsive public policy?"[24] More recently, Kevles has written a detailed companion account of the scientific and policy arrogance of some European and American biological scientists with regard to the nature of "race" and related social policy matters.[25] Egalitarianism, of course, can be overdone, perhaps especially in regard to science, but it is now very much a force for scientists to take into account.

Other cases where scientists have been denounced for imposing their views on the public are the controversy over fluoridation of the public water supplies and the controversy over the construction of more nuclear power plants.[26] In the fluoridation controversy, public distrust came from the political right in the name of the value of liberty. In the nuclear power case, distrust began on the political left in the name of peace and human welfare but has now spread to all parts of the political spectrum in the name of safety. In all these cases there is a charge of scientific arrogance and authoritarianism. In her study of the controversy over science textbooks, especially those in biology and anthropology, Dorothy Nelkin says that critics of these textbooks charge science with "an authoritarian ideology" and they express "extraordinary resentment of 'scientific dogmatism,' of the 'arrogance' and 'absence of humility' among scientists."[27] As Lipset and Schneider's data show, the public distrusts great power. The process of earning trust for themselves in an egalitarian and democratic society will be an endless one for scientists; they cannot afford to be carried away by their own means, by their good intentions, and by inattention to the public good.

## Notes

An earlier version of this paper was prepared for a symposium in honor of the late Professor Joseph Ben-David. I am grateful to the following trusted advisers for their technical competence and fiduciary responsibility: Elinor Barber, Jonathan Cole, and Viviana Zelizer.

1. For some recent attempts to overcome these difficulties, see Barber, *The Logic and Limits of Trust* (New Brunswick, N.J.: Rutgers University Press, 1983); Talcott Parsons, *Politics and Social Structure* (New York: Free Press, 1969); Niklas Luhman, *Trust and Power* (New York: John Wiley, 1980).
2. Bernard Barber, John J. Lally, Julia Loughlin Makarushka, and Daniel Sullivan, *Research on Human Subjects: Problems of Social Control in Medical Experimentation* (New York: Russell Sage Foundation, 1973), pp. 125–26.
3. There is now a large literature on these trust problems and even a field of writing and (too little) research called "bioethics." For an early and still useful survey, see Jay Katz, ed., *Experimentation with Human Subjects* (New York: Russell Sage Foundation, 1972). For an account of the "Tuskegee scandal," see James H.

Jones, *Bad Blood* (New York: Free Press 1981). And for an account of ethical trustworthiness among social scientists, see Paul Davidson Reynolds, *Ethical Dilemmas and Social Science Research* (San Francisco, Calif.: Jossey Bass, 1979).

4. Robert K. Merton, *The Sociology of Science* (Chicago, Ill.: University of Chicago Press, 1973), chap. 13, originally published 1942; See also Barber, *Science and the Social Order* (Glencoe, Ill.: Free Press, 1952).

5. For a variety of views on these technical norms by philosophers of science, see James Robert Brown, ed., *Scientific Rationality: The Sociological Turn* (Dordrecht: D. Ridel, 1984).

6. Joseph Ben-David and Terry Nichols Clark, eds., "Organization, Social Control, and Cognitive Change in Science," in *Culture and Its Creators: Essays in Honor of Edward Shils* (Chicago, Ill.: University of Chicago Press, 1977), chap. 11. See also Joseph Ben-David, *The Scientist's Role in Society* (Englewood Cliffs, N.J.: Prentice-Hall, 1971).

7. Thomas S. Kuhn, *The Structure of Scientific Revolutions* (Chicago, Ill.: University of Chicago Press, 1962); Kuhn, *The Essential Tension* (Chicago, Ill.: University of Chicago Press, 1977); Imre Lakatos and Alan Musgrave, eds., *Criticism and the Growth of Knowledge* (Cambridge, England: Cambridge University Press, 1970).

8. Harriet Zuckerman, "Deviant Behavior and Social Control in Science," in Edward Sagarin, ed., *Deviance and Social Change* (Beverly Hills, Calif.: Sage Publications, 1977); Zuckerman, "Norms and Deviant Behavior in Science," *Science, Technology and Human Values* 9, no. 1 (1984): 7–13.

9. For a detailed account of a disreputable error in a potentially important discovery, the case of "polywater," see Harriet Zuckerman, "Deviant Behavior and Social Control in Science," in Edward Sagarin, ed., *Deviance and Social Change* (Beverly Hills, Calif.: Sage Publications, 1977), pp. 111–12.

10. Robert K. Merton, *The Sociology of Science*, chap. 14.

11. For a case of competition and secrecy between laboratories headed by very able men, see Bruno Latour and Steve Woolgar, *Laboratory Life: The Social Construction of Scientific Facts* (Beverly Hills, Calif.: Sage Publications, 1979).

12. Charles Babbage, *Reflections on the Decline of Science in England and on Some of Its Causes* (New York: Scholarly Press, 1976), pp. 177–82. (Published originally in 1830.)

13. *New York Times*, May 30, 1985, p. A1; also *Science* 228 (1985): 1292–94.

14. Ibid.

15. A summary of this case with references to the several books and articles about it have been published can be found in Zuckerman, "Deviant Behavior and Social Control in Science," pp. 114–15.

16. *New York Times*, July 13, 1985.

17. In this section I am following and extending Bernard Barber, *The Logic and Limits of Trust* (New Brunswick, N.J.: Rutgers University Press, 1983), chap. 7; Barber, *Science and the Social Order* (Glencoe, Ill.: Free Press, 1952), also discussed "the social responsibilities of scientists," primarily, then, of the physicists.

18. Derek J. de Solla Price, *Science since Babylon* (New Haven, Conn.: Yale University Press, 1961); Price, *Little Science, Big Science* (New York: Columbia University Press, 1963).

19. Seymour Martin Lipset and William Schneider, *The Confidence Gap* (New York: Free Press, 1983).

20. Ibid., p. 385.

21. Dorothy Nelkin, *Science Textbook Controversies and the Politics of Equal Time* (Cambridge, Mass.: MIT Press, 1977); Nelkin, ed., *Controversy: Politics of Technical Decisions* (Beverly Hills, Calif.: Sage Publications, 1979a); Nelkin, "Scientific Knowledge, Public Policy, and Democracy: A Review Essay," *Knowledge: Creation, Diffusion, Utilization* 1 (1979b): 106–22; Nelkin, "Public Participation in Technological Decisions: Reality or Grand Illusion?" *Technology Reveiw*, (August–September) (1979c): 55–64.
22. Bernard Barber, John J. Lally, Julia Loughlin Makarushka, and Daniel Sullivan, *Research on Human Subjects: Problems of Social Control in Medical Experimentation*. For a generalized sociological discussion of scientific means becoming ends, see Robert K. Merton, *The Sociology of Science*, pp. 261–62.
23. Daniel J. Kevles, *The Physicists: The History of a Scientific Community in Modern America* (New York: A. A. Knopf, 1978).
24. Ibid., p. 425.
25. Daniel J. Kevles, *In the Name of Eugenics: Genetics and the Uses of Human Heredity* (New York: A. A. Knopf, 1984).
26. On the fluoridation controversy, see Allan Mazur, "Disputes between Experts," *Minerva* 11 (1973): 243–262; Mazur, "Opposition to Technological Innovations," *Minerva* 13 (1975): 58–81.
27. Dorothy Nelkin, *Science Textbook Controversies and the Politics of Equal Time*, pp. 131, 138.

# Part III

# The Dilemma of
# Science and Therapy

# Introduction

In 1967, under the sponsorship of the Russell Sage Foundation, I published *Drugs and Society*, a comprehensive sociological treatise on both the "dangerous" and the therapeutic drugs.[1] Resulting from my long-standing interests in the sociology of science, especially as it applied to biomedicine, and in the sociology of the professions, the book contained chapters on "Discovery and Testing Processes," "Education and Communication Processes," "Professional Specialists: Their Functions and Problems," "Some Social Problems Connected with the Use of Drugs," and "The Definition and Functions of Drugs."

In the chapter on professional specialists I included a section on "Ethical and Legal Responsibility for Experimentation on Human Beings."[2] After surveying and analyzing what was known about these responsibilties, I suggested the need for good data on what the nature and dimensions of the ethical problem of human experimentation really were. Fortunately, the Russell Sage Foundation was willing to continue its sponsorship and, with some very able junior colleagues, I undertook two empirical studies, one on a national sample of human-subject review committees and another on about 350 clinical investigators who themselves used human subjects.[3] Doing these studies, with an explicit theoretical focus on what I called "the dilemma of science and therapy," was the realization of a hope I had had to do such theoretically focused empirical research when I went to Columbia University in 1952. The last chapter of *Research on Human Subjects*, "The Social Responsibilities of a Powerful Profession: Some Suggestions for Policy Change and Reform," was an explicit statement of my concern for the application to social policy of our book's analysis and findings.

Although I did not realize it at the time, I was about to be launched on a minicareer as a policy expert on the new social problem of the ethics of research on human subjects.[4] The papers printed here in part II are some of the results of that minicareer. The papers are printed not in the order in which they were written and published but, rather, in a didactic order, which gives a better sense of the nature of this problem and of the responsibilities and regulation of the professions in general.

The first paper (chap. 8), "Medical Technology and Its Ethical Consequences," previously unpublished, was written for a conference on Modernization and Technological Innovation: Social and Economic Consequences, held at the Frank Lloyd Wright House of the Johnson Foundation in Racine, Wisconsin, in 1986. I was the arranger of the conference, at a request of the Committee on Scholarly Communication with the People's Republic of China of the National Academy of Science made in response to a proposal from the Chinese Academy of Social Sciences for a joint conference on this topic. In preparation for the conference, I visited China to consult with the Chinese Academy representatives and also lectured on medical ethics at Beijing Medical College. One of the Chinese delegates to the conference was Dr. Qui Ren Zong, now an internationally known specialist on medical ethics, which is now a worldwide social problem.

In the paper, I point out how technology has always been humankind's fortune and its fate. Until recently, most technological effects were in the physical realm. But now, biological and medical technology is having large effects. This new medical technology, together with changes throughout the social system, for example in the communications media, in increased claims to social equality, and in a heightened moral sensitivity, have together created new public concerns for improved medical ethics. Medical ethics is no longer provincially professional; it is public and political.

In the paper, eight new but still evolving social mechanisms and processes for coping with this new public and political concern for medical ethics are briefly described: national commissions of investigation, biomedical ethics research and policy institutes, publications, peer review, courts of common law, professional association codes of ethics, ethics teaching in medical schools, and the renaissance of moral philosophy.

It should be noted that the social change represented by these new mechanisms and processes has been slow and hard and is still incomplete. It has been achieved by mavericks within the medical research profession and by concerned outsiders of various kinds, over the frequent resistance of the medical research establishment. The article concludes by saying that, because of present and likely future innovations in medical technology, problems of medical ethics and ways of dealing with them will be present in modern society for a long time to come.

The second paper (chap. 9), "Perspectives on Medical Ethics and Social Change," was written as the introduction to a volume, *Medical Ethics and Social Change*, edited by me for an issue of *The Annals*, the publication of the American Academy of Political and Social Science.[5] The introduction was written as guide to some of the patterns, causes, agents, modes, resistance to, and costs of the several changes in diverse areas and aspects of medical ethics discussed in the eleven detailed and expert essays I had commissioned for the volume.

The third article (chap. 10), "The Ethics of Experimentation with Human Subjects," published in *Scientific American*, is a summary of the Barber, Lally, Makarushka, and Sullivan book, *Research on Human Subjects*, mentioned above as the report on our two studies of the ethical state of research on human subjects. Because of the very large public attention given to articles published in *Scientific American*, I was especially pleased to have this opportunity to publicize our work and our policy recommendations.

In my minicareer as an expert on biomedical research ethics, I was frequently asked by various specialist medical professional organizations to discuss the problems special to their activities. The fourth paper (chap. 11), prepared for *Reproductive Biology and Contraceptive Development: A Review of Research and Support* but never published, is one such paper. While there are certain general ethical problems in all kinds of medical research and clinical practice, there are also variations among the problems connected with neonates, the aged, and experimental subjects of any age. I was interested to discover from physicians in China, when I was there, that they were more interested in ethical problems involving neonates and the aged than in those involving experimental subjects.

The last three papers presented in part III move increasingly from the theme of the dilemma of science and therapy for the biomedical research profession to the more general theme of control and responsibility in all the powerful professions. Indeed, the very last paper (chap. 14) was written as a direct and general statement of just that theme. The fifth paper (chap. 12), "Research on Research on Human Subjects: Problems of Access to a Powerful Profession," explores one aspect of the general problem of professional power and responsibility. Using our own research experiences, it asks how the sociologist of science finds access to the powerful professions he would like to study and how he copes with the resistance to such access put up by the professions. Objective scrutiny by outsiders such as sociologists of science can be dangerous. As sociologists of science try to get "close in" to the work of active, competitive scientists, they may find access harder to achieve. The sixth paper (chap. 13), "Liberalism Stops at the Laboratory Door," points out a special irony in the access problem, namely, that scientists who are "liberal" on many social issues become conservative when it is their own practices, interests, and autonomy that are being observed. As a result they may bar access to the potentially dangerous outsider. The sociologist of science thus experiences "resistance by scientists to scientific research."

## Notes

1. Barber, *Drugs and Society* (New York: Russell Sage Foundation, 1967).
2. Ibid., pp. 89–101. A version of this section was printed as Bernard Barber, "Ex-

perimenting with Humans," *The Public Interest* 6 (Winter 1967): 91–102.
3. Bernard Barber, John J. Lally, Julia Loughlin Makarushka, and Daniel Sullivan, *Research on Human Subjects: Problems of Social Control in Medical Experimentation* (New York: Russell Sage Foundation, 1973). A summary of these studies is presented in chap. 10 of the present book.
4. For an account of some of my career activities, see my essay, "The Ethics of the Use of Human Subjects in Biomedical Research (The Prototype Case)," in Bernard Barber, *Effective Social Science: Eight Cases in Economics, Political Science, and Sociology* (New York: Russell Sage Foundation, 1987).
5. *The Annals* 437 (1978): 2–7.

# 8

# Medical Technology and
# Its Ethical Consequences

Since the very beginnings of humankind, technology has been its fortune and its fate. Throughout history, and even before, we learn, as archeology recreates humankind's earliest times, technology has had powerful effects for good and ill on social structure, culture, and morals. As humankind has learned to cope with its physical, biological, and social environments, it has done so through technological innovation; and then it has had to learn how to cope with the technological innovation itself, how to use and control it. Many of the effects of technological innovation have been unforeseen, unintended. Technological innovation seems always to require further technological innovation. From chipped flint tools to iron tools to basket-making and pottery-making to boats and ships to wheels and weapons to compasses and machines, technology has had continuous, pervasive, and powerful effects on the history of humankind.

As the few examples I have given show, until recently proportionally the biggest effects of technological innovation have been in the physical realm, with large consequences for modes of transportation, communication, and industry. I say proportionally because innovation in the physical realm always has had to be accompanied to some extent by facilitating innovations in the biological and social realms. If people are to use their techniques of mastering the physical environment, they also need to know ways of hunting, cooking, and preserving food; they need to know how to maintain their health, and they need to know tolerably effective ways of ordering and controlling their own society and of dealing with other societies. So technological innovation in the biological and social spheres has always gone along to some degree with that in the physical sphere. Still, it is in the latter sphere that technological change and its consequences have been largest and most attended to.

But now, in our time, we have come into a new world. While physical technology still continues, biological technology flourishes at about the same level as physical technology, and social technology struggles to raise its level of effectiveness to meet the possibilities and new problems created by its two sister technologies. The great burst of biomedical technology in our time starts slowly and sporadically in the nineteenth century with such discoveries as anesthetics and the stethoscope. In the twentieth century, and especially in the last fifty years, we have seen the development of antibiotics, oxygen tents, heart-lung machines, CAT scanners and magnetic resonance scanners, the computers and computer programs to make these scanners and other medical technology possible, endoscopes of various kinds such as laryngoscopes and colonoscopes, artificial hearts, kidney dialysis machines, and techniques for amniocentesis.

All of this new biomedical technology has had powerful effects on physicians, their patients, the costs of medical care, and the ethical problems faced by physicians, patients, and those responsible for the decisions on how to allocate and pay for the improved medical care and longer life this new technology makes possible. By creating and using the new biomedical technology, we have also created a new set of ethical problems: When does life begin, at conception or at some later point of the viability of the fetus? By what equitable social mechanisms should access to the benefits of the new medical technology be distributed and controlled? Do physicians as clinical experimenters with new medical technology have the right to use human subjects without their informed consent or to expose them to unfavorable risk-benefit ratios in such experiments? How shall we give a new definition to death when heart-lung machines make it possible to keep brain-dead patients "alive"? Who has the right in which circumstances to decide when a terminally ill patient shall die: the patient, the family of the patient, the attending physician, the hospital staff, a hospital ethics committee, or a judge in the court system?

These are the central ethical questions, posed in many different and specific forms, that now are of general ethical concern in societies that possess some or all of the new biomedical technology. But we should emphasize that it is not the new biomedical technology alone that has caused the new types and level of ethical concern. The technology interacts with a number of other social and cultural changes that have occurred in modern society to create, together, the new atmosphere of ethical awareness and concern.

Among these changes we may mention, first, the great increase in communications about medical matters in general and about medical technology in particular. These are now matters of "front-page," "prime-time" interest to the media of communication because they are of urgent interest to their readers and viewers and listeners. All the media of communication—newspapers, television, radio, magazines—now have specialized staff to cover and report the latest news in this field. A striking example is the way in which news of medical discoveries in the *New England Journal of Medicine (NEJM)*, perhaps the premier American general medical journal, is handled by the media. The *NEJM* is published each Thursday. The media are not allowed to report its contents before that day. But media staff may inspect the contents on Wednesday and send any important news to their employers for publication on Thursday, in the early editions of newspapers and radio news, on prime-time evening television news, if important enough. The immediacy of this reporting means that the general public may read about, hear about, or see news of a medical discovery before the copy of the *NEJM* reaches a physician subscriber through the mail. And we know that medical news in the media is closely attended to by the public. Studies show that, of all the news in the media that can be said to be "scientific," the news coming from the medical field is far and away of the greatest interest to the public. The media have been essential, of course, in creating medical "scandals," that is, in describing medical situations or the use of medical knowledge in a way that the investigative reporters who flourish in these times in media reporting condemn as shameful, unethical, or inequitable.

Thus it was a television investigative reporter who revealed and "created" the scandal about the retarded institutionalized children at Willowbrook in the New York area who were being used as experimental subjects, without the informed consent of their parents, in research on the causes of infectious hepatitis. It was another New York newspaper reporter who disclosed the "scandal" of the injection of live cancer cells into geriatric patients, again without their informed consent, by a distinguished local cancer researcher. And, finally, it was a newspaper reporter for a southern newspaper who reported that black men in a Tuskegee, Alabama, population in which syphillis was endemic had been made the subjects of an experimental study without their informed consent and had been "scandalously" denied treatment with antibiotics when those became the treatment of choice after World War II. This latter "scandal" had large effects in bringing national legislative attention to the problems of the ethical use of human subjects in medical experiments. As part of this legislative attention, during hearings before both Senate and House of Representative committees, my colleagues and I were able to offer research evidence from two studies we had done, one of a national sample of medical institutions using human subjects and the other of a sample of some 300 medical researchers who had used human subjects, that there was indeed a general problem of unethical use of such subjects. Following on these hearings, Congress created the President's Commission for the Study of Ethical Problems in Medical and Biomedical and Behavioral Research. We shall look a little further into the work of this commission later on. So, the media, and the "scandals" they can create in the public mind, are important parts of the communications processes that nowadays raise ethical concern in addition to the biomedical technology itself.

The importance of these media "scandals" points to a second cultural factor that heightens ethical concern with the consequences of biomedical technology. These medical scandals do not occur in a moral vacuum. I believe that there is a heightened moral sensitivity, indeed even sometimes an undesirable moral dogmatism and absolutism, on which media reporters seeking to create "scandals" can play in modern society. This moral sensitivity is more aware of and concerned about the weak, the powerless, the poor, the disadvantaged, and also about those positions, persons, and processes that treat abusively or corruptly or unfairly those in need of support and special help. Fetuses, the terminally ill, the aged, the poor, and ignorant subjects of medical experiments, even (through anthropomorphic extension) experimental animals, are now the objects of heightened moral sensitivity.

Finally, increased egalitarian sentiments in the modern world interact with new technology and increased communications and heightened moral sensitivity to raise the intensity and level of concern about the consequences of technological innovation in the biomedical field. The American Hospital Association has formulated, and many of its hospital members now post, a Patient's Bill of Rights that specifies new egalitarian and equitable claims that patients may make in the formerly all-powerful and somewhat impersonal atmosphere of the modern hospital with its omnipresent, invasive medical technologies. Egalitarianism and increased moral sensitivity combine to give strong support to the assertion by ethicists, patients, and physicians that all patients, rich and poor alike, should be treated the same, with dignity.

In sum, there is now so widespread and persistent a public and professional concern with the ethical consequences of biomedical innovation that a whole new research, policy, and personal-advice profession of bioethics has emerged to deal with this concern. When I describe, below, the several new social mechanisms that have been created to bring attention to and devise solutions for these new ethical consequences of biomedical technology, I shall have more to say about various aspects of this new bioethics movement and profession.

It is essential to note the new *public* character of medical ethics. As Berlant has shown in his account of the history of medical ethics, until recently medical ethics was an intra-professional concern, more a matter of medical etiquette, of how physicians were to deal with one another, than of larger public morality. There was little public involvement in defining the public goals and the specific substance of medical morality. This has now changed under the influence of the powerful consequences of the new medical technology and the associated social and cultural changes that I have described above. Medical ethics has become, in the largest sense of that term, politicized; it is of essential concern to the general public and its legislatures, government agencies, and courts of law. It is still a professional matter, of course, but, because of its politicization, professional ethics is no longer merely a matter of ethics but one of larger public morality. Professional medical ethics is too important to be left to the professionals alone. They are now required to make their codes acceptable to the public and its political agencies.

Given this public, political character of medical ethics, what social mechanisms and processes have already emerged and are still evolving to produce a more satisfactory moral response to the ethical problems that the new medical technology has created?  I should like to describe a number of them briefly, and not in any order of their importance; they should also be seen as interrelated in many ways.

## 1. National Commissions

As a result of the Tuskegee scandal and subsequent committee hearings occasioned by this scandal in the Senate and in the House of Representatives of the United States Congress, legislation was passed establishing in 1976 the President's Commission for the Study of Ethical Problems in Medicine and Biomedical and Behavioral Research. The members of the commission were chosen to be broadly representative of the general public and of the relevant professional and expert groups. There were members of broad segments of America's pluralistic society: physicians, of course; lawyers; social scientists, and philosopher-ethicists of diverse religious affiliations. Unlike some government commissions, this one was not designed to cover up problems. It worked actively and over sufficient time, first to define the nature of the new ethical problems brought about by new medical technology and then to propose a series of general moral principles and specific moral practices to realize those principles.  The commission was well funded and employed a large staff of experts of various kinds: lawyers, social scientists, moral philosophers. The commission and its staff solicited expert judgments from expert outsider consultants and funded public-opinion research on matters relevant to its own views and decisions. The outcome was a considerable number of volumes reporting the commission's findings and recommendations, which, in line with the commission's very general mandate, were comprehensive, covering every aspect of the new medical problems from birth to death. The commission's activities were continuously and widely reported in the media. The commission gave not only a larger visibility but a new legitimacy to the new problems of medical ethics. The commission's recommendations did not result in new legislation but they did affect both the government administrative agencies responsible for medical ethics and the courts of law before whom cases involving specific ethical decisions came for adjudication.

For example, in the administrative areas, in response to the commission's *First Biennial Report* (1982), the Office of Science and Technology Policy in the Executive Office of the President set up the Ad Hoc Committee for the Protection of Human Research Subjects and, subsequently, the Interagency Human Subjects Coordinating Committee, composed of representatives of twenty different federal agencies. On the basis of the deliberations of these two groups, the Office of Science and Technology Policy formulated (1986) a Model Federal Policy for the Protection of Human Subjects. This is administrative and not statutory regulation.

## 2. Biomedical Research and Policy Institutes

Happily for the field of medical ethics, its new and complex problems have come to be closely and continuously attended to by specialized research and policy institutes, which have taken on themselves the intellectual and moral responsibility for this field. The pioneer institute of this kind is the Hastings Institute, named for the suburban village near New York City in which it is located. Founded in 1970, the Hastings Institute is a free-standing organization, that is, not connected with the government or any university or corporation. It is a prime example of that essential voluntarism that De Tocqueville saw, 150 years ago, as an essential characteristic of American society. It was founded by Dr. Daniel Callahan, a philosopher-ethicist, and Dr. Willard Gaylin, a psychiatrist-ethicist, who have succeeded in creating, on their own energy and initiative, an important social resource. The institute has not, until recently, had any endowment and has depended on research and conference support from a wide variety of foundations and corporations and from the membership fees of a large number of interested citizens. The Board of Directors and the full-time staff, headed by Callahan and Gaylin, have been the essential components of the institute, but they have been aided by a variety of special consultants, advisers, and expert ad-hoc committees. There is no problem of medical ethics on which the institute, during its existence, has not established a study committee to investigate and make policy recommendations. The institute has had an important educational function. Its fellowship program, with Fellows staying and working for up to a year, has trained cadres with expert knowledge to go back to universities, government, and other institutes. In addition to this professional educational commitment, the institute has had a continuing and important general public educational function. The institute is nearly universally known for its activities and policy recommendations to the mass media, both locally in its metropolitan area of New York and nationally. As a result, whenever some medical ethical problem arises in the daily news, editors call the Hastings Institute for its views. The philosophers, social scientists, and lawyers on the staff of the institute, often having already studied at least the general aspects of the relevant problem, are able to give expert and considered judgment on relatively short notice. Such contributions to the public discourse can very much improve its quality.

The Kennedy Institute of Bio-Ethics at Georgetown University is another important scholarly and policy forum for the field of medical ethics. Georgetown is a Catholic university and its Kennedy Institute takes its inspiration from that religious affiliation in at least a loose sense. Its members are members of the Georgetown University faculty, from its philosophy and theology departments. Like the Hastings Institute, the Kennedy Institute members are frequently consulted by the media, by government agencies, and by medical groups. Some of its members served the President's Commission as members, staff, or consultants.

## 3. Publications

The bioethics movement has produced a flood of valuable publications: research studies; encyclopedias; anthologies; textbooks; case studies; and journal articles in specialized journals, law reviews, medical journals, and general magazines; and, finally, a few publications devoted entirely to medical ethics. The *Hastings Center Report* has probably been the most influential of these, with excellent articles and wide circulation and readership. Another valuable publication has been *IRB*, edited by Dr. Robert Levine at Yale University Medical School and devoted to helping IRBs (institutional review boards), the ethical peer review committees that now exist at all institutions doing biomedical and behavioral research, to become more effective. *IRB* contains articles about such matters as informed consent, staffing and record-keeping by IRBs, and various substantive problems newly confronting IRBs.

## 4. Peer Review

IRBs and other peer review mechanisms are one of the most important social-technology innovations of the bioethics movement. By the early 1960s a few medical schools had apparently set up not too effective peer review committees to review the ethics of experiments using human subjects. The big turning point came in 1966, when the National Institutes of Health (N.I.H.) began to require all institutions that it funded to have such committees. The N.I.H. acted, it has been reported by Dr. Donald Frederickson, then a high official and later director of the N.I.H., and now the president of the Hughes Medical Institute, out of fear of legal liability. The N.I.H. lawyers convinced its physician administrators that such liability could be avoided by the peer review arrangement. Since practically every institution in the United States in the 1960s was receiving a considerable part of its funding from the N.I.H., there was no delay, although some grumbling, in complying with the N.I.H. mandate. Peer review committees were established everywhere and have now become a fixed part of the biomedical and behavioral research world. The original prescription of the N.I.H. *recommended*, but did not require, that a few members of the IRBs be lay persons. Usually these were chosen from among lawyers, clergymen of various denominations, and community activists, all people with some interest in biomedical ethics and with some disposable time. More recently, such lay members have been *required* of IRBs. The N.I.H. will not make a grant for any research proposal that has not been ethically approved by an IRB. No satisfactory research study of the IRBs has been done, but they seem to have been effective not only in eliminating gross ethical delinquencies in regard to informed consent and satisfactory risk-benefit ratios but in heightening the awareness of researchers in these matters. Knowing the ethical review was going to occur, researchers have regulated themselves in advance. Government regulation of the ethics of biomedical research is present in the sense that the IRB review is required, but the substantive regulation is essentially left to the biomedical profession-als themselves, with some participation by the laypersons on the IRBs. The N.I.H. maintains a small Office for Protection from Research Risks to over-see the IRBs but it has only a small staff and no considerable activities.

The peer review modality has spread beyond its N.I.H. mandate. Acting partly out of moral concern and partly out of fear of legal liability, hospitals have set up ethical review committees to adjudicate ethical problems, particu-larly in connection with defective births and problems of dying and death. Again, we have here a useful device for internal self-regulation in the medical world.

## 5. Courts of Law

But peer review and ethical review committees have not been the final answer to the need for ethical control of the problems brought about by new medical technology. Partly because of the indecision or fear of these commit-tees themselves and partly because their decisions and actions have some-times been challenged by colleagues, hospital staff, or members of families of particular defective births or dying patients, the ethical decisions of the com-mittees have been brought to the courts for the last word. It should be made clear that there is very little federal or state legislation in American society that covers medical ethics. The controls that exist come from administrative agencies such as the National Institutes of Health and from the common-law courts, usually those of the several states, not the federal courts. In a halting and not altogether integrated way, as is the case with all common-law procedure, these courts have been feeling their way toward a set of principles and rules for dealing with medical-ethical problems that have never before come to their jurisdiction. Perhaps the prototypical and certainly the most highly publicized case to come before a state court was the case of Karen Quinlan, in which compassionate parents sought to remove life-support systems from their brain-dead daughter. Similar and somewhat different cases in the areas of defective children and dying persons have come to the courts since. Gradually a set of precedents in the common-law style is being established but it is far from settled and the courts continue to have to make decisions in ethical matters where there is no precedent and, presumably, both law and ethics are being created.

## 6. Professional Association Codes of Ethics

Because of the increased visibility and politicization of the new problems of medical ethics caused by innovations in medical technology, several powerful medical professional associations have had to take the new moral and political climate into account. They have responded by reworking their existing codes of ethics to include principles applying to the new ethical problems of informed consent, defective births, and dying. The altered codes are useful guides to general principles but they are no substitute for the specific rules and practices that are the responsibility of the IRBs, other ethical review groups, and the courts of law. The professional associations have not yet set up their own effective regulatory committees. On the whole, as establishment organizations, they have followed and not led in the bioethics reform movement.

## 7. Medical Ethics Teaching

Recognizing that teaching medical ethics to their students might be a partially effective preventive measure against the ethical delinquencies and uncertainty that required later regulatory procedures in actual medical and hospital practice, U.S. medical schools have set up, just about universally, a wide variety of programs for the teaching of medical ethics. Required and elective courses, seminars, and lectures have all been used to heighten general awareness in this field and to provide some specific guidance. No reliable research has been done on the effectiveness of these teaching programs. Impressionistic evidence, however, suggests that they are not taken too seriously. On the whole, American medical schools are still dominated by the values of scientific medicine to the relative neglect of social and nonscientific value concerns. Students quickly discern the dominant values of the curriculum and of their instructors and tend only to put up with, rather than commit themselves seriously to, such ancillary courses, as they are subtly defined, as medical ethics. Not until medical students are recruited for more than scientific aptitude, not until curricula are changed, and not until the scientist-stars of the medical faculties are matched in their influence by their maverick colleagues who place great stress on the new problems of medical ethics, will medical students learn their ethical lessons in the way they learn the scientific ones.

## 8. The Renaissance of Moral Philosophy

For reasons that have not yet been sociologically investigated, there has been a renaissance of moral philosophy in the American academic community roughly from the mid-1960s on. This renaissance has succeeded the previous emphasis on problems of logic and language in the academic philosophy community. This renaissance has been an essential component of the bioethical movement in two respects. First, at the intellectual level the renaissance has aroused renewed concern for the moral problems of justice, equality, human dignity, and individual autonomy. Second, at the occupational level, because of the recent decline of growth in American universities, newly trained Ph.D.s have been unable to secure academic positions. Many of them have been attracted to the alternative opportunities offered by the expansion of bioethical concerns. Medical schools and hospitals have become large employers of philosophers willing to specialize in the moral problems of medical ethics. To the scientists who run medical schools and hospitals, concern for values has seemed to be special province of philosophers and other humanities. The world of bioethics, as a result, has now been flooded with and come to be dominated by philosophers who are imbued with the new interest of their professional field in moral problems and who are looking for jobs now no longer available in universities and colleges. Up to a point, this influence of the philosophers on the bioethics movement has been valuable. But their influence has not been sufficiently supplemented by the perspectives and research practices of lawyers, sociologists, and political scientists. Some of these have participated in the bioethics movement, but medical scientists prefer philosophers.

The eight social processes and mechanisms described above as emerging and evolving for more effective coping with the new problems of medical ethics caused by innovations in medical technology and associated factors represent a considerable social change. Most of this change has occurred since the mid-1960s. But it has not occurred entirely easily. Joined by a small minority of professional mavericks, that is, those willing to challenge the professional establishment in its set ethical ways, outsiders to the medical profession have been chiefly responsible for the changes that have occurred on moral rather than legal liability grounds. Philosophers, theologians, social scientists, journalists, and members of Congress have led the way in exposing ethical problems and in proposing and carrying out reforms. Professional vested interests, especially in what has been defined as the necessary and sometimes seemingly absolute autonomy of medical scientists, have been the source of considerable resistance to the changes that have occurred. Because of this resistance and because social change is often slow and hard, there is still room for improvements in the social control mechanisms for medical ethics. Professional self-regulatory agencies need to be strengthened, medical practitioners need to be earlier and better educated to the new ethical dilemmas, and a more extensive body of common-law precedents needs to be accumulated. Still, no matter how one looks at the medical-ethics scene today, the glass has to be seen as at least half full, and perhaps more so. A great deal has been achieved in these years.

Nevertheless, despite this achievement, it is clear to us now that more remains to be done in coping with the new medical technology. Not only is there room for improvement because of past innovations, but new innovations are already appearing and it is likely that the flow of innovative medical technology will continue. New medical technology based on the rapid progress being made in molecular biology and in its derivative biotechnology, for example, has already raised ethical problems of the most fundamental kind. Will such new technology result in undesirable human engineering? Will it create the potential for fundamental changes in the human gene pool? Who will control such engineering and gene-pool changes? Such questions have been raised and they challenge our ethical and social resources. We shall be living with the problems of medical ethics caused by innovations in biomedical technology for a long time to come.

## Selected Bibliography

Babbie, Earl R. *Science and Morality in Medicine*. Berkeley, Calif.: University of California Press, 1970.

Barber, Bernard. *Informal Consent in Medical Therapy and Research*. New Brunswick, N.J.: Rutgers University Press, 1980.

———. *The Logic and Limits of Trust*. New Brunswick, N.J.: Rutgers University Press, 1983.

———, ed. *Medical Ethics and Social Change*: Annals of the American Academy of Political and Social Science, vol. 437 (May 1978).

———, John J. Lally, Julia Loughlin Makarushka, and Daniel Sullivan. *Research on Human Subjects: Problems of Social Control in Medical Experimentation*. New York: Russell Sage Foundation, 1973.

Beauchamp, Tom L., and James F. Childress. *Principles of Biomedical Ethics*. New York: Oxford University Press, 1979.

——, and LeRoy Walters, eds. *Contemporary Issues in Bioethics*. Encino, Calif.: Dickenson, 1978.

Brim, Orville G., Jr., Howard E. Freeman, Sol Levine, and Norman A. Scotch, eds. *The Dying Patient*. New York: Russell Sage Foundation, 1970.

Crane, Diana. *Social Aspects of the Prolongation of Life*. New York: Russell Sage Foundation, 1969.

Freund, Paul A. *Experimentation with Human Subjects*. London: George Allen & Unwin, 1972.

Goodfield, June. *Playing God: Genetic Engineering and the Manipulation of Life*. New York: Random House, 1977.

Gray, Bradford, H. *Human Subjects in Medical Experimentation*. New York: John Wiley, 1975.

Jones, James H. *Bad Blood: The Tuskegee Syphilis Experiment*. New York: Free Press, 1981.

Katz, Jay. *The Silent World of Doctor and Patient*. New York: Free Press, 1984.

——, ed. *Experimentation with Human Beings*. New York: Russell Sage Foundation, 1972.

——, and Alexander Morgan Capron. *Catastrophic Diseases: Who Decides What?* New York: Russell Sage Foundation, 1975.

Levine, Robert J. *Ethics and Regulation of Clinical Research*. Baltimore, Md.: Urban and Schwarzenberg, 1981.

President's Commission for the Study of Ethical Problems in Medicine and Biomedical and Behavioral Research. *Reports*: (1) *Defining Death* (July 1981); (2) *Making Health Care Decisions* (October 1982); (3) *Splicing Life* (November 1982). Washington, D.C.: U.S. Government Printing Office.

Reich, Warren T., ed. *Encyclopedia of Ethics*, 4 vols. New York: Free Press, 1978.

Reynolds, Paul Davidson. *Ethics and Social Science Research*. Englewood Cliffs, N.J.: Prentice-Hall, 1982.

Rivlin, Alice M., and P. Michael Timpane, eds. *Ethical and Legal Issues of Social Experimentation*. Washington D.C.: Brookings Institution, 1975.

Shannon, Thomas A., ed. *Bioethics*, rev. ed. Ramsey, N.J.: Paulist Press, 1981.

Titmuss, Richard M. *The Gift Relationship: From Human Blood to Social Policy*. New York: Pantheon Books, 1971.

# 9

# Perspectives on Medical Ethics
# and Social Change

Medical ethics are now in a period of great change. The world of medical science and the society around it are now both undergoing great changes, and it is these that inevitably cause the change that is occurring in medical ethics. Not that changes in medical ethics have not occurred before. Berlant's essay on the history of medical ethics describes and explains sociologically some of the changes that have taken place in the past, from Hippocrates forward. But the change occuring now is of a scope and rate that marks the present period as a very special one in the history of medicine. For some laypersons, the present changes are not fast enough; for some physicians and physician-investigators, they are all too fast; for everyone, there is no question that something of the first importance and novelty is now happening in medical ethics.

### Some Causes of Change

At first glance, the great changes that have occurred recently in basic and applied medical sciences might seem sufficient cause of the changes that are occurring in medical ethics. On this premise, one might argue that the great new control that biomedical advances have given us over morbidity and health, life and death, might be a sufficient cause of the correlated changes in medical ethics. But this technological determinism will not do. Two other causes, working together with scientific changes, are necessary for a better explanation of the causes of the present changes in medical ethics. These other causes are changes now going on in two areas of values and interests that are of fundamental importance in our society. One of these areas of value and interest has to do with rationality; the other, with egalitarianism.

Rationality is a powerful set of values and interests in our society that predisposes people to search out the most efficient and effective means to their ends. With increasing education and increasing assimilation of the population into our fundamental values has come an increasing emphasis on rationality in all spheres of behavior. For ends so important to people as health and life, it is obvious that rationality commands people to get as much good medical care as they can. They know how much good it can do them; but they have also come to know that it can do them ill, that medical enthusiasms can sometimes bring harm, that people can be used as subjects in medical experimentation that is of no immediate benefit to them, and that they may lose some of their autonomy and control over their ways of living and dying. Medicine, they have learned, has diverse consequences, for both good and ill, and they rationally prefer more of the good and less of the ill. They want medical ethics changed to maximize the good the doctors and physician-investigators do, minimize, without ever totally eliminating, the harm they also cause, however inadvertently.

Egalitarianism is another powerful set of values and interests in our society. It predisposes people to want to increase their autonomy, to reduce somewhat the consequential differences of resources and prestige between themselves and their fellows, to have somewhat more say in the vital decisions affecting their welfare that powerful authorities and elites may be making without their participation and consent. In our society, and elsewhere in the world, we see blacks as against whites, women as against men, young as against old, clients as against professionals, and patients as against doctors, asking for somewhat less authoritarianism, even somewhat less paternalistic benevolence, somewhat more equality. Equality is on the march against overbearing expertise, and one of our key problems is how to accommodate these two important sets of values and interests to each other more satisfactorily. One of the areas where we seek this more satisfactory accommodation between competing values is medical ethics.

## Some Change Agents

Whatever its diverse and interrelated causes, social change in any area is not embodied in wholly impersonal mechanisms but operates through the actions of a variety of individuals who serve as agents of change. In the area of medical ethics, these change agents have been some insider medical professionals, a miscellaneous set of "humanists" or "bioethicists," some social scientists who have done policy-relevant research, and some people from government.

Insider change agents have been people like the late Professor Henry Beecher of the Harvard Medical School, whose 1966 article in the *New England Journal of Medicine* describing some twenty-five articles in the professional medical and research journals that showed evidence of unethical use of human subjects, stunned his colleagues into an awareness of the problem.[1] Unfortunately, while Beecher raised the level of consciousness of the medical research world, his analysis was too individualistic, too psychologistic. He thought he was dealing just with "bad guys." What others have pointed out is that it was a "bad system" that was turning "good guys" into "bad guys."[2] Another important set of insider professional change agents has been an anonymous group of researchers and administrators at the National Institutes of Health (N.I.H.) who, since the early 1960s, have labored to construct effective regulations for local peer review of the ethical aspects of all biomedical research funded by them. Their great achievement has been creation of the 1966 N.I.H. ethical regulations. Without Beecher and the N.I.H. professionals, recent changes in the medical ethics of human experimentation would have been less effective and less speedy. In other areas of change in medical ethics, such as problems of abortion, genetic counseling, and humane death, insider professionals have been equally important.

But insiders, as we shall see shortly, have been more resistant than receptive to these changes. Change would not have occurred without the set of outside change agents from the fields of ethics, philosophy, the law, and the humanities, who have come to be known, inclusively, as "bioethicists."[3] These change agents have held conferences, written books and papers, given lectures, held seminars, and provided training facilities for all those newly interested in medical ethics and have even established specialized institutes to consider its problems. The leading institutions investigating medical ethics are the Kennedy Institute of Bio-Ethics at Georgetown University and the Hastings Institute, a private enterprise headed by Daniel Callahan and Willard Gaylin. The bioethicists have come from those fields where a general professional concern for moral values is important. Their involvement with medical ethics is only the latest, and certainly not the last, of the expressions of their general moral concern.

Another set of outsiders has been the professional social scientists who have carried out careful research on problems such as human experimentation and the treatment of dying patients, and who have been much concerned for formulating better medical ethics and better social policy on the basis of the knowledge acquired in their research.[4] These outsiders have felt that the availability of essential facts is indispensable for the rational processes of change to affect medical ethics. The kinds of abstract, unspecified, and unsupported statements, which are often forthcoming both from the insider professionals and from the outsider bioethicists are not adequate for successful change in ethics and policy.

Finally, some government people have been important change agents. Senator Jacob Javits of New York was responsible for inserting ethical-review clauses into the Kefauver-Harris Drug Amendments of 1962. Senator Edward

Kennedy of Massachusetts and his aids and advisers have had a long-standing, continuing, and key influence on the legislation for improving medical ethics, especially in connection with the legislation in 1975 for setting up the National Commission for the Protection of the Human Subjects of Biomedical and Behavioral Research, an agency that has provided an excellent model of how new regulations for medical ethics may be constructed with due regard to all the values and interests involved. Beyond congressmen, of course, there has been the important influence, referred to above, of the medical researchers working for the government in the National Institutes of Health. The combination of their professional concerns and the power of the National Institutes, as government funding agencies, has made their effects on change extensive and beneficient indeed.

## Modes of Change

Important social change, such as is now occurring in medical ethics, does not proceed along a single path but through multiple and diverse modalities. One such modality is what sociologists call a "social movement." A social movement consists of a diverse, often diffuse (partly organized, partly not) aggregation of people and forces that push toward some new social goal. It is often more effective in raising a moral cry than in defining specific ways of achieving its goal, more successful in raising the level of social consciousness in the public domain about some new social evil than in carrying through the actual process of practical reform. The various people and forces pushing for change in medical ethics have something of this character of a social movement. Their diverse, diffuse, overlapping, somewhat conflicting energies are necessary for the birth of the idea of reform, but they are not enough. Moral change has to be supported, established, supplemented, and sometimes led by governmental and legal action.

Governmental and legal action is the other important change modality that has been so significant in causing important changes in medical ethics. But such action, if entirely unsupported by a strong moral base, would be ineffective; it can lead morality a little bit but can never get too far ahead of it. In the case of medical ethics, fortunately, the moral changes created by a social movement have been successfully implemented by a great variety of governmental and legal changes in many different areas of medical ethics. Since the reltionships between social movements and governmental and legal actions, between morality and law, are of great consequence in all areas of social change, a close study of what has happened in the area of medical ethics should be instructive for those interested in change in other areas of society.

## Resistance to Change

On the whole, physicians and physician-investigators have been resistant to the changes presently occurring in medical ethics. They have been conservative in the literal sense, wishing to preserve the status quo. This is not surprising. All social groups tend to be conservative when they feel their central values and interests are being challenged and subverted, and physicians and physician-investigators now feel that their important values and interests, with regard to both autonomy and expertise, are being challenged and subverted.

What the doctors and the researchers see as threats to their valued autonomy, of course, are seen by the reformers in medical ethics as necessary changes in effective regulation of powerful professionals, who either will not effectively regulate themselves or who have seized too much power and need the influence that laypersons have a right to assert in the name of their own values and interests. Again, it is not surprising that physicians and physician-investigators resist some reduction in the great autonomy that is presently theirs and, in considerable measure, accepted as legitimate by the public. We live in a society where there is a constant tension between the claims for autonomy by powerful professions (and other social groups) and the demands for more effective governmental regulation and/or other forms of regulation by the public vitally affected by the established autonomies. What we see happening in the area of medical ethics happens elsewhere in the society: processes of conflict and resolution between competing claims for autonomy and demands for regulation. What is needed, of course, in medical ethics as elsewhere, is less fruitless conflict and more rational accommodation and resolution. Despite a good deal of persisting indifference, hostility, and resistance from those who oppose changes in medical ethics, on the whole rational accommodation prevails and results in the kinds of changes in medical ethics that the different parties at interest can live with.

In our society, the two values of equality and expertise are in constant tension in many different areas, not just in medical ethics. Demands for changes in medical ethics come from many sources that seek to redress the balance of inequality that now gives too much authority, some call it "authoritarianism," to the medical profession. The claims of egalitarianism assert themselves in requests for laypersons to sit on ethical peer review boards, to allow the individual to decide for himself or herself how to die, to obtain kinds of information that a reasonable, prudent person needs in order to give his or her informed consent to a medical procedure. The claims of medical expertise assert themselves in statements that only the expert can really understand what is going on in medical relationships and medical research. Obviously, this is an overreaching on the part of the experts, just as there is often overreaching on the part of egalitarians; for example, when they reject entirely the medical claim to some technical scientific knowledge and skill. (The laetrile controversy and gynecological self-treatment are cases in point.) Overreaching from either side is not what we need to work out new and effective accommodations between equality and expertise in the area of medical ethics. We need knowledge-based resolutions that will satisfy both laypersons and powerful expert professionals that their respective values and interests are being recognized, respected, and maximized so far as is possible in the inevitable and unending tension between the two camps.

### Some Social Costs of Change

Social change never proceeds without some social costs beyond the benefits that also are produced. Such costs occur for all the parties involved. Just now, for example, because changes in medical ethics are not occurring as rapidly as they would like, many laypersons are dissatisfied with medical professionals. There is much "murmuring of the masses," much expression of informal dissatisfaction, in addition to formal protests. On the other side, physicians and physician-investigators complain about the decline in trust and public confidence in their efforts. They view themselves as the benefactors and fiduciaries of the public's interests. They are not only affected in their pocketbooks by medical malpractice suits that inflate their malpractice insurance costs (after all, these can often be passed on to patients and third-party payers; physicians are hardly low-income people), they are morally outraged by attacks on their integrity and moral worth. On the other side, those laypersons who enter suits for malpractice are similarly morally outraged, not only by the hurt they have suffered but also by the lack of effective self-regulation in the profession that would provide an arena, other than the civil courts, for airing their grievances.

There are social costs of other kinds. On the one side, many laypersons find it hard to act in the more independent way the new medical ethics recognizes as their right; they have been trained to dependence, deferentiality, and submissiveness, and they cannot change. They also find it hard to participate effectively in the new mechanisms of regulation that give some place to laypersons. What evidence we have, and it is not much, shows that "community" members on medical-affairs boards are still very much dominated by the authoritative professionals who sit with them. On the other side, professionals brought up in the old modes of inequality and expert authority are impatient with, and resentful of, the changes required in their ingrained ways of feeling and behaving. This is what we mean when we say that social change is not easy to come by, that it is hard and slow. On both sides, or often on the many sides of a changeful situation, the old ways are hard for people to abandon. Change has its strains and its social costs to all parties.

## Outlook for the Future

Strong and conflicting moral values surround many of the areas of medical ethics presently in process of change. When this is true, we cannot expect instantaneous or final resolutions of many of the existing tensions and conflicts. Medical ethics will continue to be changing for some time to come. It is clear, of course, that much beneficial change has already occurred; the last few years have been a remarkable period in the history of medical ethics. But as the processes of change continue, we hope with less conflict and more rational remedy, we will need all the dispassionate social analysis and objective empirical facts that we can get to help us along.

## Notes

1. Henry K. Beecher, "Consent in Clinical Experimentation: Myth and Reality," *New England Journal of Medicine* 195 (1966): 34–35.
2. John J. Lally and Bernard Barber, "'The Compassionate Physician': Frequency and Social Determinants of Physician-Investigator Concern for Human Subjects," *Social Forces* 53 (1974): 289–96. See also Bernard Barber, "Compassion in Medicine: Toward New Definitions and New Institutions," *New England Journal of Medicine* 295 (1976): 939–43.
3. For a discussion of bioethics, see Renée C. Fox, "Advanced Medical Technology — Social and Ethical Implications," *Annual Revue of Sociology* (1976), pp. 231–68.
4. Diana Crane, *The Sanctity of Social Life: Physicians' Treatment of Critically Ill Patients* (New York: Russell Sage Foundation, 1975); Bradford H. Gray, *Human Subjects in Medical Experimentation* (New York: Wiley-Interscience, 1975); Bernard Barber et al., *Research on Human Subjects* (New York: Russell Sage Foundation, 1973).

# 10

# The Ethics of Experimentation with Human Subjects

The power, scope, and funding of biomedical research have expanded enormously in the past forty years. So also, inevitably, has clinical research with human subjects. That expansion has led to widespread reflection on what is increasingly perceived as a new social problem: the abuse of human subjects of medical experimentation. In particular it is alleged that human subjects are not always protected from undue risk and do not always have the opportunity voluntarily to give their adequately informed consent to participation in experiments.

A social problem is defined in part by the concern it arouses, and this one has clearly aroused concern. Members of the medical profession itself led the way, with increasing numbers of journal articles, books, and seminars on the issues. The public has become aroused, largely through popular accounts of dramatic incidents—genuine scandals in certain cases—involving the violation of the dignity and rights of patients. And the federal government has moved to protect human subjects, potential or actual. Beginning in 1966 the National Institutes of Health, the Food and Drug Administration, and the Department of Health, Education, and Welfare have issued increasingly detailed regulations governing experimentation with human subjects in projects they support, which means in most of the biomedical research done in the country. In 1974 a National Commission for the Protection of Human Subjects of Biomedical and Behavioral Research was established to advise the Department of Health, Education, and Welfare, and it has been replaced by a long-term National Advisory Council that is to deal with the same issues.

The regulations, commissions and councils and the very fact of interference in medical activities by outsiders are viewed by many investigators as being onerous and ever dangerous. On the other hand, many outsiders believe far more social control is required. The debate on the issue has been conducted without much reference to objective evidence. In 1970, my Research Group on Human Experimentation undertook two studies of investigators' attitudes and practices. On the basis of our results I would argue that there is indeed inadequate ethical concern among biomedical investigators, that it is reflected in excessively risky procedures, and that better internal and external controls are essential.

There are two major reasons for the general recognition that experimentation with humans is a subject for concern, one of which I alluded to at the outset: the increased power, scope, and funding of biomedical research. The other reason is a change in values: increased emphasis on equality, participation, and the challenging of arbitrary authority.

It is easy to forget how new scientific medicine is. The revolutionary advances based on knowledge of physiology, and biochemistry have come since the 1930s, and they came from research. The basic work could be done with test-tube preparations and laboratory animals, but eventually human subjects had to be involved. Man is "the final test site," as Henry K. Beecher, a pioneer among physicians concerned about the ethics of research, once put it. Unfortunately there are no statistics on the number of people who are subjects in medical experiments or even on how many projects involve human subjects; the National Institutes of Health (N.I.H.) keeps records according to area of research (a disease or a physiological process, for example) rather than according to species of experimental subject; the N.I.H. can say only that recently about a third of the projects it approves involve human subjects. It is clear, however, that the number of human subjects is larger than it used to be and that some small but significant minority of those subjects are involved in risky experiments. If more people have been put at more risk, then there is a rational basis for concern about the satisfactory balancing of risks and benefits, about adequate protection from unnecessary risk and about some groups being put at more risk than other groups.

Over and beyond this utilitarian basis for the new social concern with medical experimentation is the value factor, which arises from recent social changes. All over the world individuals have been demanding more equality of treatment and the right to be informed about and to participate in decisions affecting them, and they have been challenging the right of experts to make those decisions unilaterally. People who define themselves as being unequal, underprivileged, or exploited are demanding better treatment and better protection, whether it is underdeveloped countries as against developed ones, blacks as against whites, women as against men, young as against old, patients as against doctors—or subjects as against investigators. This moral revolution of rising value-expectations has combined with the revolution in medicine to focus attention on the ethics of experimentation with human subjects.

45. A researcher plans to study bone metabolism in children suffering from a serious bone disease. He intends to determine the degree of appropriation of calcium into the bone by using radioactive calcium. In order to make an adequate comparison, he intends to use some healthy children as controls, and he plans to obtain the consent of the parents of both groups of children after explaining to them the nature and purposes of the investigation and the short and long-term risks to their children. Evidence from animals and earlier studies in humans indicates that the size of the radioactive dose to be administered here would only very slightly (say, by 5–10 chances in a million) increase the probability of the subjects involved contracting leukemia or experiencing other problems in the long run. While there are no definitive data as yet on the incidence of leukemia in children, a number of doctors and statistical sources indicate that the rate is about 250/million in persons under 18 years of age. Assume for the purpose of this question that the incidence of the bone disease being discussed is about the same as that for leukemia in children under 18 years of age. The investigation, if successful, would add greatly to medical knowledge regarding this particular bone disease, but the administration of the radioactive calcium would not be of immediate therapeutic benefit for either group of children. The results of the investigation may, however, eventually benefit the group of children suffering from the bone disease. Please assume for the purposes of this question that there is no other method that would produce the data the researcher desires. The researcher is known to be highly competent in this area.

45A. Hypothetically assuming that you constitute an institutional review "committee of one," and that the proposed investigation has never been done before, please check the *lowest* probability that *you* would consider acceptable for *your* approval of the proposed investigation. (Check only *one*)

( ) 1. If the chances are 1 in 10 that the investigation will lead to an important medical discovery.
( ) 2. If the chances are 3 in 10 that the investigation will lead to an important medical discovery.
( ) 3. If the chances are 5 in 10 that the investigation will lead to an important medical discovery.
( ) 4. If the chances are 7 in 10 that the investigation will lead to an important medical discovery.
( ) 5. If the chances are 9 in 10 that the investigation will lead to an important medical discovery.
( ) 6. Place a check here if you feel that, as the proposal stands, the researcher should not attempt the investigation, no matter what the probability that an important medical discovery will result. (*IF YOU CHECKED HERE*, please explain):_____

45B. Which of the above responses comes closest to what you feel the *existing institutional review committee* in your institution would make? _____ (Please write in the number of the response.)

45C. Which of the above responses comes closest to what you feel the *majority of the researchers* in your institution would make, acting in their role as researcher rather than as a "committee of one"? _____ (Please write in the number of the response.)

10.1 HYPOTHETICAL EXPERIMENT described here was one of six experiments submitted to investigators and administrators on hospitals and other research centers in a mailed questionnaire. In each case respondents were asked whether, under specified conditions, they would approve of the experiment. This proposal involved giving radioactive calcium to children with a bone disease and to a control group and measuring its uptake by bone.

Public awareness of the problem is too much the result of headlined scandals, but the scandals do illustrate some of the possible abuses. In the 1960s two respected cancer investigators who were studying the immune response to malignancies injected live cancer cells into a number of geriatric patients at the Jewish Hospital and Medical Center of Brooklyn without first obtaining the patients' informed consent. A few years later a leading virologist conducted an experiment at Willowbrook, a New York State institution for the severely retarded. Reasoning that a serious liver infection, hepatitis, was in effect endemic in the hospital anyway, he deliberately exposed some children to hepatitis virus in an attempt to achieve controlled conditions for testing a vaccine. The accusation was that the children's parents were not given enough information on which to base informed consent, and that in some cases consent was given perfunctorily by administrators of the institution.

More recently there was the exposure by the press of the ongoing syphilis experiment in Tuskegee, Alabama. Since the 1930s a group of black subjects with syphilis had been kept under observation in an effort to study the course of the disease. That was not considered wrong in the 1930s, when the known treatments for the disease were only marginally effective, but by 1945 penicillin had become available as a safe and extremely effective cure for syphilis. Yet somehow the experiment was continued, and presumably some men died of the disease who could have been cured.

How significant are such scandals? We do not know, because no one has been doing the kind of social bookkeeping about numbers of subjects, degree of risk, adequacy of consent, and efficacy of protective mechanisms that would yield an overall view of experimentation with human beings and that might contradict the more extreme allegations of abuse elicited by the publicized scandals. In the absence of such intensive record-keeping it remains for social research to fill the gap by sampling the total range of experimentation with human subjects. To that end the Research Group on Human Experimentation conducted first a national mail survey of nearly 300 biomedical research institutions and then an intensive interview study of 350 individual investigators at two institutions.

The national survey questionnaire was answered by 293 teaching and non-teaching hospitals and other research institutions that, our analysis showed, constituted a nationally representative sample of all such institutions. Those who filled out the questionnaire were generally themselves active researchers and members of their institution's review committee, set up to pass on research proposals. We asked the investigators to give us their response to six simulated proposals such as those that might come before a review committee. The proposals were detailed research protocols designed to measure the degree of the investigators' concern about informed consent and their willingness to approve of studies involving various levels of risk. We could be confi-

44. It has been shown that the thymus has an important bearing on the development and maintenance of immunity. For this reason the researcher proposes an investigation to determine the effect of thymus removal on the survival of tissue transplants, a very timely and important problem. In a sample of children and adolescents admitted for surgery to correct congenital heart lesions, he would randomly select an experimental group for thymectomy. Though the thymectomy will prolong the heart surgery by a few minutes, there is otherwise extremely little additional surgical risk from this procedure. At the conclusion of each heart operation, a full-thickness skin graft, approximately one cm. in diameter and obtained from an unrelated adult donor, would be sutured in place on the chest wall of both the experimental and control groups. He would then compare the survival of the skin grafts in each of the groups. It has been shown in a number of investigations of neonatal rats and other animals that those whose thymus had been removed were much less likely to reject skin grafts. The possible long-term immunological problems that might result are as yet not completely know, but a number of studies in animals indicate significant, immunological deficiencies after thymectomy. Studies done in humans with myasthenia gravis, some of whom had undergone thymectomy, have not definitively demonstrated that the immunological abnormalities discovered in these patients were the result of thymectomies. To quote one authority: "There were no immunologic abnormalities that could be attributed to the effect of thymectomy *per se*."

The research will result in no therapeutic benefits for the patients involved. The researcher plans to obtain the consent of his potential patient-volunteers and/or their parents after explaining the procedures involved in the investigation as well as the possible short-term surgical and long-term immunological hazards for the subjects.

10.2 REMOVAL OF THYMUS GLAND during heart surgery was the experimental procedure proposed in another protocol in the questionnaire. Respondents were asked if they would approve of the experiment, given various probabilities that it would show thymectomy "considerably increases the probability of tissue-transplant survival in children and adolescents."

dent that the protocols were "hypothetical-actual" rather than "hypothetical-fantastic" because we constructed them with careful attention to the research literature, checked them with specialists and pretested them with a dozen chiefs of research at medical centers, who found them to be convincingly real.

One protocol described a study of chromosome breakage in users of hallucinogenic drugs; blood samples (for chromosomes) and urine samples (for evidence of drug use) were to be taken, at no risk but also without notification of the experimental purpose, from students routinely visiting the university health center. Another protocol proposed that the thymus gland, which is a component of the immune system, be removed unnecessarily from a random sample of children undergoing heart surgery; the objective was to learn the effect of the thymectomy on the survival of an experimental skin graft made at the same time. The other protocols dealt with a random test of alternative treatments for a congenital heart defect in children; with an evaluation of the efficacy of a new drug for severe depression (placebos were given to some patients); with a study of lung function in patients kept under unnecessarily prolonged anesthesia after undergoing a routine hernia repair, and with an investigation of the effect of radioactive calcium on bone metabolism in children [*see illustrations* 10.1 *and* 10.2].

The answers to the thymectomy, anesthesia, and radioactive-calcium protocols in particular gave us measures of the respondents' attitudes toward the balancing of risks and benefits. A clear pattern emerged. In the case of the high-risk thymectomy, for example, 72 percent of the respondents said the project should not be approved no matter how high the probability was that it would establish the efficacy of thymectomy in promoting transplant survival. On the other hand, 28 percent of the respondents said they would approve the experiment; 6 percent said they would approve it even if the chance of significant results was no better than 1 in 10. Similarly, 54 percent were against doing the calcium study at all—but 14 percent said they would approve it even if the odds were only 1 in 10 that it would lead to an important medical discovery. Our basic finding was that whereas the majority of the investigators were what we called "strict" with regard to balancing risks against benefits, a significant minority were "permissive," that is, they were much more willing to accept an unsatisfactory risk-benefit ratio.

The same general pattern of a strict majority and a permissive minority emerged from our second study, in which we interviewed 350 investigators actively engaged in research with human subjects. The investigators were at institutions to which we gave the synthetic names University Hospital and Research Center, and Community and Teaching Hospital. The institutions were picked (by a technique known as cluster analysis) as being representative of two kinds of medical centers that do considerable amounts of research. The interviewees told us about 424 different studies involving human subjects, and for each study they estimated the risk for subjects, the potential benefit for subjects, the potential benefit for future patients and the potential scientific importance of the study. It was reassuring to find that the investigators considered that only 56 percent of the clinical investigations graded for risk and benefits involved any risk for the subjects. We went on, however, to cross-tabulate the estimated risks and benefits and we concluded that in 18 percent of the studies the risk was not adequately counterbalanced by the benefits. We called those studies the "less favorable" ones, and we proceeded to classify them further according to their potential benefits for other patients or for medical science. Even when these compensating justifications were taken into account, tabulation revealed a "least favorable" category of studies in which the poor immediate risk-benefit ratio was not compensated for by possible future benefits. These "least favorable" investigations constituted 8 percent of the investigations in our analysis.

The concept of informed consent is a troublesome one. The investigator wants to have enough subjects and is afraid of scaring them off. Patients are likely to to be concerned about their own condition, may feel powerless with respect to the physician or hospital, and often have difficulty understanding medical language or concepts. Even established medical procedures can have somewhat unpredictable consequences, so that physicians feel there is a limit to how completely "informed" a patient can be. The fact remains that regulations of government funding agencies and most institutions now require that the human subject of an experiment (or his guardian, in the case of small children and mentally incompetent patients) understand that something is being done (or some treatment is being withheld) for reasons other than immediate therapeutic ones; the subject or guardian must be informed of any risks and must give consent voluntarily.

With regard to informed consent, our questionnaires and interviews again revealed a minority with "permissive" views and practices, although that minority was smaller than it was for unfavorable risk-benefit ratios. For example, 23 percent of the questionnaire respondents said they would approve the chromosome-break proposal, which presented the informed-consent issue clearly in effect by itself. The situation was more complex in the heart-defect protocol. Here other dubious elements competed with the fact that the investigator would not inform the parents that his decision whether or not to operate would be a random one, not based on therapeutic considerations. Only 12 percent of our respondents said they would approve of the study without requiring any revisions, but only 65 percent specifically mentioned the lack of informed consent as a problem.

The best available research evidence on informed consent comes from a study conducted by Bradford H. Gray, who was then a graduate student at Yale University, at a distinguished university hospital and research center (not the one in our interview study). With the consent of the responsible investigator, Gray interviewed 51 women who were the subjects in a study of the effects of a new labor-inducing drug. Although the women had signed a consent form, often in the hectic course of the admitting procedure or in the labor room itself, 20 of them (39 percent) learned only from Gray's interview, which was held after the drug infusion had been started or even after the delivery, that they were the subjects of research. Among those who did know, most of them did not understand at least one aspect of the study: that there might be hazards, that it was a double-blind experiment, that they would be subjected to special monitoring and test procedures or that they were not required to participate; four of the women said they would have refused to participate if they had known there was any choice. Many of the women had been referred for the study by their private physician, but instead of being informed that an experimental drug was to be administered they were told that it would be a "new" drug; they trusted their doctor and assumed that "new" meant "better."

How does it happen that the treatment of human subjects is sometimes less than ethical, even in some of the most respected university-hospital centers? We think the abuses can be traced to defects in the training of physicians and in the screening and monitoring of research by review committees, and also to a fundamental tension between investigation and therapy. We have data bearing on each of these causative factors.

It is in medical school that the profession's central and most serious concerns are presumably given time and place and that its basic knowledge and values are instilled. Yet the evidence from our interviews shows that there is not much training in research ethics in medical school. Of the more than 300 investigators who responded to questions in this area, only 13 percent reported they had been exposed in medical school to part of a course, a seminar, or even a single lecture devoted to the ethical issues involved in experimentation with human subjects; only one respondent said he had taken an entire course dealing with the issues. Another 13 percent reported that the subject had come to their attention when, as students, they did practice procedures on one another; for 24 percent it was in the course of experiments with animals; 34 percent remembered discussion of ethical issues in specific research projects. One or more of these learning experiences were reported by 43 percent of the respondents—but the remaining 57 percent reported not a single such experience. The figures were about the same whether the investigators were graduates of elite U.S. medical schools, other U.S. schools, or foreign schools. The figures were a little better, however, for those who had graduated since 1950 than for older investigators.

What little ethics training there is is apparently not very effective: the investigators who reported having learned something about research ethics were only slightly less permissive in response to protocols presenting the risk-benefit issue than those who reported no such experiences. It would appear that both the amount and the quality of medical-school training in the ethics of research could be improved. In this connection it is worth remembering that the many physicians who are not engaged in investigation at all also need some background in experimentation ethics, if only so they can evaluate requests that they direct their patients toward a colleague's research project.

Scientific "peer review" is a keystone of scientific inquiry, operating implicitly in many ways and explicitly in the case of professional journals, grant-awarding committees, and many institutional reviewing boards such as the "tissue committees" that assess the results of surgery in hospitals. Ethical peer review of experimentation with human beings should be the counterpart of scientific peer review, but until the mid-1960s such activity received limited support among biomedical researchers. Even after 1966, when the N.I.H. mandated ethical peer review for all its grantees, effective review did not become universal. Our questionnaire went to hospitals and other research centers that had filed with the N.I.H. formal assurances that the required institutional review committee had been established, but 10 percent of the respondents said their institution's committee reviewed only proposals for outside funds and 5 percent reported that only formal proposals to the N.I.H. were reviewed. The two institutions in our interview study were among the 85 percent that stated they were reviewing all research proposals, and yet 8 percent of our interviewees volunteered the information that at least one of their own investigations with human subjects had not been reviewed.

How effective are the review committees in handling the protocols that do come before them? Our questionnaire respondents told us that in 34 percent of the institutions the committees had never required any revisions, rejected any proposals, or had any proposals withdrawn in anticipation of rejection for ethical reasons; 31 percent reported revisions, 32 percent outright rejections, and 19 percent withdrawals. Either some of these committees have very few ethical problems coming before them or they are ineffective. Gray's study in an institution with an active and strong committee suggests that they are ineffective rather than underworked. The committee whose performance he examined found relatively few proposals that did not need some kind of modification, and he thinks "a record of few actions by committees is an indication that their members are indifferent or that their standards are loose."

The peer-review groups seemed weak in other ways. In some institutions there was no face-to-face discussion among the reviewers. Only 22 percent of the committees had members from outside the institution, something that was then recommended and has since been mandated by the Department of Health, Education, and Welfare. In practically none of the institutions was there continuous monitoring of studies that were approved, although this was even then required by government regulations. In general, ethical peer review is hampered by the fact that each committee operates in isolation and must consider every new issue on its own and without benefit of precedent. A case-reporting system, such as operates in the law, would make that unneccessary and would promote both equity among institutions and high standards. The major weakness in the system is the lack of keen interest in and support of the review committees on the part of most working biomedical investigators. Research is their business; research is their mission and predominant interest, not applied ethics or active advocacy of patients' rights.

Most biomedical investigators are, however, interested in taking care of patients and making them well. As a result medical institutions and individual investigators operate today with two powerful sets of values and goals. On the one hand, there is the pursuit and advancement of scientific knowledge. On the other, there is the provision of humane and effective therapy for patients. Through a broad range of complex interactions these two sets of values and goals are harmonious, even complementary and mutually reinforcing. Occasionally, however, scientific research and humane therapy can be in conflict. When that happens, there is sometimes a tendency to choose the pursuit of knowledge at the expense of the ethical treatment of patients. An irreducible minimum of conflict may be inevitable. The ethical task now is to come as close as possible to that minimum—and to resolve unavoidable conflict in favor of humane therapy.

There is evidence that the enhanced excitement attending scientific achievement and the rewards bestowed on it in recent decades have skewed the decision-making process in many cases of conflict. As our data show, the medical schools have been largely indifferent to training their students in the ethics of research. Moreover, their record in peer review has been inferior to that of other institutions. Answers to our questionnaire showed they were less likely than other research centers to have set up a review committee before the N.I.H. required one, less likely to have one that met the first N.I.H. guidelines in 1966, less likely to have a committee that reviews all clinical research, and less likely to include on their committee medical or nonmedical members from outside the institution. Medical schools, the Association of American Medical Colleges, and professional associations of clinical investigators have been much quicker to seek research funds or to protest funding cuts than to organize seriously for the purpose of studying the ethics of research and making policy in that area.

The same emphasis on the pursuit of knowledge rather than on ethics is apparent among individual biomedical investigators. Ethical concern for the subjects of their research is not a major factor when they select their collaborators; at least it is not often mentioned as a characteristic they look for in collaborators. Scientific ability is a major concern. When we asked our 350 interview subjects, "What three characteristics do you most want to know about another researcher before entering into a collaborative relationship with him?" 86 percent of the respondents mentioned scientific ability, 45 percent mentioned motivation to work hard and 43 percent mentioned personality. Only 6 percent of them listed anything we could classify as "ethical concern for research subjects."

The tension between investigation and ethical concern is perhaps best illustrated by indications that the struggle for scientific priority and recognition exerts pressure on ethical considerations. Our data show that the social structure of competition and reward is one of the sources of permissive behavior in experimentation with human subjects; the relatively unsuccessful scientist, striving for recognition, was most likely to be permissive both in his approval of hypothetical protocols and in his own investigative work. We divided our respondents into four categories based on the number of papers they had published and the number of times their work had been cited by other workers; the frequency of citation has been shown to be a good measure of scientific excellence. We called the most-cited investigators the "high-quality" scientists and those who had published a great deal but were never cited the "extreme mass-producer" scientists. It was the extreme mass producers who were most often engaged in investigations with less favorable risk-benfit ratios, who approved of the protocols with poorer risk-benefit ratios, and who least often expressed awareness of the importance of consent. Caught up in the socially structured competitive system of science, unsuccessful in it but still pursuing the prize of peer recognition, they appear to be more likely to overvalue scientific work as against humane therapy.

It is not only the mass producers, contending for recognition among peers in their discipline, who are apt to be more permissive. We also weighed the rank achieved by each worker within his own institution against various measures of his effectiveness compared with that of his colleagues. We found that the "under-rewarded" investigators tended to be the more permissive. There is also a quite different kind of medical investigator who, we think, is likely to be pushed toward permissive practices by scientific competition: some of the professionally esteemed, highly successful medical scientists who are engaged in intense competition for priority and recognition in well-publicized areas of research. There are not many of those people, and they did not emerge in our sample, although some workers who refused to be interviewed may belong in that category. In the absence of real data we can only point to such evidence as published discussions concerning the worldwide heart-transplant competition of a few years ago, which raised questions about the premature exposure of human subjects to what were then still experimental procedures.

Given the fact that there are ethical defects in current medical research standards and practices, do the resulting abuses strike particularly, as is often alleged, at certain social groups: at the poor, at children, and at institutionalized patients (prisoners in particular)?

The evidence from our interviews with 350 investigators indicates that the poorer patients in hospitals are indeed at a disadvantage as subjects of research. For each of the 424 studies our respondents reported, they told us whether fewer than 50 percent, between 50 and 75 percent, or more than 75 percent of the subjects were ward or clinic patients (as opposed to patients in private or semiprivate rooms and under the care of their own physician). We found first of all that ward and clinic patients were more likely to be subjects of experiments. Moreover, when we examined the cases we had previously identified as having "less favorable" and "least favorable" risk-benefit ratios, we found that both categories were almost twice as likely to involve subjects more than three-quarters of whom were ward and clinic patients as the studies with the more favorable ratios were.

The ward and clinic patients are, of course, vulnerable to that kind of discrimination. They can most readily be channeled into an experimental group by admitting physicians and clerks without interference from a personal physician. They tend to be less knowledgeable about hospitals, more readily intimidated, and less likely to understand what they are told about an experimental project, and therefore less likely to be able to withhold their consent or to give genuinely informed consent. In sum, they are the least likely to be able to protect themselves.

Many institutionalized patients are poor and perhaps incompetent, and they may feel completely dependent on the institution's administrators and physicians. Prisoners are a special case: they are institutionalized in an implicitly coercive situation, so that genuinely informed consent may be a logical impossibility. On the other hand, a prison population is by definition a good source of experimental and control subjects living under controllable conditions, and there have been instances where prison studies have been conducted humanely, with good scientific results and apparently with good effect on the prisoners' morale. Experimentation with prisoners is nevertheless subject to grave abuses. For instance, the head of the Food and Drug Administration told a Senate committee that a review of experimentation in nineteen prisons revealed abuses ranging from unprofessional supervision of drug tests to inadequate medical care and follow-up treatment.

Children constitute still another special group. Small children cannot give consent for their own participation in experiments; older children, who could, are often not asked. As the Willowbrook incident demonstrated, parents are not always adequately protective of their children's interests. In the case of institutionalized patients, prisoners, and children, new regulations of the Department of Health, Education, and Welfare call for special protective committees and procedures. These will be effective, however, only in a context of better ethical training for investigators and more effective peer review.

The ethical problems that attend medical research with human subjects are representative of an entire class of problems created by the impact of professionals and professional power on the general public and on public policy. In the area of research with human subjects, the medical investigators are not alone; there is a tendency in other fields, too, for humane concerns to be left at the laboratory door. Psychologists and sociologists have often been accused of circumventing the requirement for consent and of applying unethical manipulative techniques in their investigations of human behavior, and neither profession has welcomed scrutiny from outsiders or restrictive regulation. The issue goes beyond research ethics, however. Many professions now command knowledge that has great potential usefulness for human welfare but bestows power that can be abused. Because professional power is largely based on knowledge that has not yet diffused to the general public it must to a considerable degree be self-regulated, but because professional power is of such major public consequence it must also be subject to significant public control. The medical-research profession does not have a proud record of self-regulation or acceptance of public controls.

# 11

# Ethical Aspects of Clinical Research in the Field of Human Reproduction

Knowledge gives power, and power mightily affects social values, for good and ill. So it has been in our time with biological knowledge in general and also with that particular part of biological knowledge that is included within reproductive biology and contraceptive development. Just as physics and chemistry began to have fundamental consequences for social values in earlier centuries, so reproductive biology in this century has come to be a prime force of the same kind. Some of our central value-concerns have been touched: the nature and sacredness of life, the meaning of kinship and sexual behavior, the privacy and autonomy of the individual, the propriety of social controls. It is no wonder, then, that the whole advance of knowledge about human reproduction and contraception is attended by numerous and fundamental ethical problems. For the moment, concern with resources for scientific advance and with its actual achievements have outrun both our concern and our capability for dealing with the ethical problems that scientific progress and achievements have helped to bring into being. There is a certain lag here, between scientific and ethical capabilities. Something is being done to reduce the lag, but not enough yet.

The ethical problems brought into being by the advance of knowledge in reproductive biology can usefully be divided into two classes. The first consists of those ethical problems arising from the use of human beings as experimental subjects in research, no matter what type of biomedical research it is. The second class of ethical problems consists of those that are peculiar to research in reproductive biology, problems such as those arising from doing abortions, using fetal tissue, doing *in vitro* fertilization, doing genetic screening and counseling that has consequences for reproductive behavior, and the like. In the first part of this essay we shall deal with the general problems of human experimentation. In the second, we shall discuss the special problems of research on the reproductive systems and behavior of human beings.

## Ethical Problems in Human Experimentation

We begin with the problems of human experimentation in general. While much research in reproductive biology limits itself to test-tube and animal work, there is, of course, much research that is and can only be done on human subjects.[1]

In rudimentary form, medical experimentation has existed since the cure of illness began. But in its systematic, pervasive, and scientific form, experimentation on human beings is one of the best products of the twentieth century. Its remarkable achievements are everywhere to see. This century will surely be marked as the time when biomedical science came to maturity, both in theory and in its application.

Since the human being is the animal of necessity in all biological research that has any proximate therapeutic intent, experimentation on human beings has increased steadily, perhaps exponentially, in this century, as we have learned more about physiological processes, scientific techniques (e.g., double-blind procedures, sampling, etc.), and about the therapeutic efficacy of such agents as antibiotics. One hears less and less in medicine of that ancient maxim, "First of all, do no harm." Increasingly, the prescription is for the heroic interventions that our progress has made possible. Scientific medicine is more and more the goal and even, happily, more and more the reality. Medical schools concentrate on instilling the scientific ideal as well as scientific practice. And a number of other well-established medical institutions join wholeheartedly with the medical schools in fostering this scientific ideal and practice.

Human experimentation is necessary and good for scientific medicine, but is it good when judged by other standards or in the light of other ends? Just because of the great increase in experimentation, with its attendant costs of inconvenience, risk, and occasional harm for many subjects, this is a question that would probably have been asked anyway, on grounds of rational calculation of interest. But it is all the more a question that has been asked in recent times because of a great moral change that characterizes our age as much as medical progress does. This is the age of the citizen—as Professor T. H. Marshall put it in his Marshall Lectures of 1949[2]—of the member of society with claims to genuine equality of dignity and hence to equality of treatment in all respects, with the right to share somehow in the making of decisions that affect his or her vital interests and values. In short, it is the age of civil rights: for black as well as white, for the young as well as the old, for women as well as men, for students as well as teachers, and—for present purposes—

for subjects as well as biomedical scientists. The American Hospital Association has recently recognized this explicitly in its Bill of Rights for Patients.[3] The bill gives formal promulgation to the new climate of opinion, for both the therapeutic and the research treatment of all hospital patients.

The profession of medical research has had a part in asking the question "Is experimentation always good for its human subjects?" Several world and national medical conferences, from Nuremberg to Helsinki, have formulated codes of ethics for the protection of human subjects. Medical men like Dr. Henry Beecher in the United States and Dr. M. M. Pappworth in England have published works[4] that demonstrate that experimentation is not, in practice, always good for its subjects, and they have appealed to research workers to change their ways. Medical scientists at the U.S. National Institutes of Health (N.I.H.) were instrumental in formulating the regulations issued in 1966, requiring all research supported by the N.I.H. to be scrutinized for its conformity with ethical standards by locally constituted committees made up of professional peers. In England, the Royal College of Physicians has recommended a voluntary system of "peer-group review." However, according to the World Health Organization, only in the United States is there a governmentally required system of control, through peer review, over experimentation. In the nongovernmental field, and specifically for the field of reproductive biology research, both the Ford Foundation and the Population Council, which are probably the largest nongovernmental sources of research funds in this area, have formulated procedures for the protection of human subjects, which all their grantees must follow. In addition, the Ford Foundation uses a consultant to scrutinize the ethical aspects of all grants using human subjects, even though ethical deficiencies are often noted earlier by the scientific referees. And finally, many of the specialized societies of the medical profession have had sessions, at their annual meetings over the last few years, on the ethics of experimentation.

And yet one cannot say that the medical profession has given to this matter the serious, intensive, continuing attention it deserves, whether one considers the medical profession's own traditional values or the new demands for better practice in this field, which the public and its representatives are now making. If one compares the initiative and activity of biomedical scientists in seeking funds for research and training with their efforts to clarify, teach, and practice ethical standards in experimentation, one finds the latter wanting. And apart from such necessary new institutional mechanisms as an effective system for reporting on the ethical problems of the profession or a commission on standards and training, which has not yet been created by the medical profession, there persists a lack of attention to these matters, and a widespread ignorance, which probably accounts for the weakness of institutional arrangements. Though there has been a good deal of improvement since the N.I.H. mandated peer review in 1966 for all research using human subjects, much remains to be done, as we shall indicate below.

Even in the face of some of the unjust charges made in connection with the Tuskegee, Alabama, syphilis experiments, biomedical research workers have been unable to make out the good case for their profession that they could have made. Biomedical research apparently still has the trust of the public in the United States, while the therapeutic side of the profession seems to be losing it—if we may judge from the rising numbers of malpractice suits. If confidence in research is to be maintained, the research profession itself will have to do more than it has done so far.

Nonetheless, the training that the American medical schools provide for biomedical research workers in the ethical problems of experimentation has been quite unsatisfactory. Only 13 percent of a sample of 350 medical research workers studied by us in two typical institutions—(1) University Hospital and Research Center, and (2) Community and Teaching Hospital—said they had had a seminar, a lecture, or part of a course devoted to the issues involved in the use of human subjects. Thirteen percent said that the ethical issues of research had come up when, as students, they performed practice procedures on one another. Twenty-four percent said they became aware of the issue when they were doing experimental work on animals. Thirty-four percent remembered discussions with teachers or with other students on the ethical issues involved in specific research projects, which they learned of in class or through the professional or scientific literature. But 57 percent reported that they had never in any way been brought into such confrontation with the ethical problems of experimentation.[5]

There is obviously a great need for more concerned and thoughtful attention to this matter. And this is a need not just for those physicians who do research and who use human subjects, but for all physicians who are likely to recommend their patients at some point to hospitals and clinics where research is part of the routine. The physician who wishes to meet his responsibility to his patient in such a situation must know something about the realities of research and of the use of human subjects. Even in the United States, where there are now mandatory mechanisms for meeting the requirements on the two key issues of informed consent and favorable risk-benefit ratio, and where the situation has certainly improved somewhat as a result of obligatory peer review, the medical-research profession still has some way to go to conform with its own values and to satisfy public demands for ethical treatment.

Another of our two studies, of a representative national sample of 292 medical research institutions that use human subjects and that had given assurance of compliance to N.I.H., showed that 86 percent of them review all research; 9 percent still do not conduct such reviews unless there is a formal research proposal; and 4 percent reported that they review only research proposals that they submit to N.I.H. for financial support. Moreover, of the 350 medical research workers who were employees in the two institutions reported on above, which ostensibly review all research, 9 percent of the respondents volunteered the information that they were doing research that had not been reviewed by peer groups for conformity with the required standard. And even where peer review has been carried out, some of it is not being done under what we roughly define as the most efficacious of conditions. For example, there is a lack of continuous review in many institutions, even though this is required by N.I.H.; there is no face-to-face discussion among the reviewers; and there is no procedure for appeal. The committees in the several institutions were not in touch with one another and hence lost the advantage that is conferred by shared experience. The medical school peer-review committees were no better than such committees in other research institutions. The biomedical research world is pervaded by an emphasis on autonomy and individualism; these have an essential place but, untempered by other values and emphases, they are costly.

What are some of these costs? One thing is clear. According to the respondents' own estimates of the benefits and risks connected with their own reserach, which comprised 424 different studies using human subjects, some investigations are still being done that involve more risk than benefit—not just to their present subjects, but even when benefits to possible future patients are considered. What they themselves told us showed that 18 percent of the studies involved more risk than benefit for present subjects; we call such investigations "less favorable." Some of these studies were said to promise benefits to future patients; we called such investigations "least favorable." Even these "least favorable" investigations made up 8 percent of the studies that were reported to us and evaluated by the researchers themselves. Since these are the researchers' own estimates, remember, it is not unlikely that at least some small underestimation of risks, some small overestimation of benefits is involved.

Our data make another thing clear about costs to subjects. This is that it is the ward and clinic patients, rather than the private patients, who are more likely to be the subjects in both the "less" and the "least favorable" investigations. These are the people least likely to understand a study in order to give informed consent. They are less knowledgeable about how hospitals are organized and about what goes on in them other than patient care. In sum, they are least likely to know how to protect themselves. This fact is inconsistent with the medical profession's traditional obligation to treat all patients equally.

This difference in the treatment of subjects of different social classes has also been observed in a study at another leading hospital and research center conducted by Professor Bradford Gray of the University of North Carolina.[6] Inquiring into the extent of "informed consent," Professor Gray discovered that only 1 out of 3 clinic patients in a study using a new drug for the induction of labor in childbirth knew, at the time of admission, that she was being used as an experimental subject. While the private patients were a little better off and knew that they were subjects in 1 out of 2 cases, half of them too had not been properly informed. Moreover, instead of being told than an "experimental" drug was being tested, they were told only that a "new" drug would be used. Trusting their private physicians, and assuming that "new" implied "better," they were ignorant that they were research subjects—until they were in the labor room or, in some cases, until they were interviewed by Gray after giving birth. All this occurred in an outstanding medical institution where peer review exists and where the medical principal investigator had encouraged Professor Gray to make his study. On the basis of his interviews with research subjects and his scrutiny of the correspondence between investigators and the peer-review committee, Gray concludes that while peer review has many uses, it does not ensure that the actual processes of obtaining informed consent are satisfactory. Important deficiencies remain in this fundamental ethical area. These things happen because research workers are ignorant of, or do not take seriously enough, their ethical responsibilities to their human subjects. Science is often more important to them than humane therapy.

This conflict between science and therapy seemed to explain some of the finding of our studies. We found that those scientists who are less successful in the international scientific community, that is, whose work was cited by other scientists relatively infrequently, were more likely to have done the "less favorable" studies. Scientists who thought that they were not properly acknowledged by their own research organizations were also more likely to have done the "less favorable" studies. In these cases, scientific ambition tended to prevail over humane concern. Of course, scientific ambition among very distinguished scientists, especially when competition between them is very close and very visible, may also become overpowering. Such cases have been reported in the press, but we encountered none of them in our interview.

There is no need to conclude from all this that biomedical scientists are "mad" or irresponsible. The tension between science and humane therapy is inherent in biomedical science but, as a result of great scientific progress during the last decades, the biomedical profession has become somewhat negligent of the ethical responsibilities which progress imposes. More attention to these responsibilties, more concern with them, is now essential if science and humane therapy are to better accommodated to each other.

There are two further aspects of the general ethical problems of experimentation on human subjects in which researchers on reproductive biology have been involved. One is related to use of so-called captive or incompetent subjects in these experiments:  prisoners, the mentally retarded, and children. The other is related to use of overseas subjects by American researchers or the use of those subjects by their fellow nationals with research subsidies from the American government or American foundations or pharmaceutical companies. Prisoners and the mentally retarded have apparently been used in reproductive research, and so also have overseas populations.

In the case of captive and mentally incompetent subjects, the charge that their present use is unethical rests on the assertion either that it is impossible in principle to get satisfactory informed consent from them, or that it is possible only with special safeguards that go beyond even those used with free and competent adult populations. While some civil liberties groups have taken the first line, the second line has recently been taken by the National Institutes of Health in a draft proposal for a possible regulation providing special safeguards for prisoners, children, and the mentally defective. This discussion proposal describes a system of special protection committees over and beyond the peer-review committees whenever any of these three classes of research subject are to be used. Such committees are not only to scrutinize the proposed consent processes with special care but to monitor the actual ongoing research and consent activities and be available to subjects for special consultation and help needed in being a research subject. Given the defects in the actual consent processes shown by Gray to exist even for adult competent populations, it was probably inevitable that such special protection arrangements would be provided first for captive and incompetent populations. The National Institutes of Health has received an unusually large amount of comment and criticism on this draft proposal from research and civil liberties groups. It is unlikely, in the face of this response, to give up the principle of special protection, but it probably will modify the specific arrangements somewhat insofar as comment can show them to be excessively cautious or excessively time-consuming. However, the fact that the new arrangements for higher ethical standards will impose costs on research personnel time and activities is not in itself an argument against them. As new ethical values and standards are formulated, as humane therapy comes to be given somewhat greater weight as a value in relationship to the value of scientific progress, some costs against the latter value are inevitable. Arguments in cost terms, therefore, are likely to have weight only insofar as they can be shown to be unnecessary for the goal in view of redressing the balance between science and humane therapy. An argument purely in cost terms denies the importance of the humane-therapy and informed-consent values.

In the case of the use of overseas subjects by American researchers or by fellow national researchers using American research funds, the charge that their use is unethical rests on wild assertions that "genocide" is being committed or in milder assertions that such research exploits people in other countries for the benefit of Americans. In the probably rare case where the benefit was entirely for future American patients, the use of overseas subjects exclusively would seem to be clearly unethical. In the more likely case that a favorable outcome of the research would benefit patients in the country at issue as well as in the United States and perhaps elsewhere in the world, the ethical standing of the research is more complicated. No matter whether the benefit is to be shared by all parties, the risks and responsibilities should be similarly shared. If they are not so shared, then the charge of exploitation by Americans will have weight. It should be carefully noted by American fund-granting agencies that any claim by them that they were only acceding to the requests for research support from overseas researchers doing allegedly unethical research is not likely to absolve them from ethical and political responsibility. It does indeed seem to be the case that ethical controls on the use of human subjects are weaker in some overseas research sites than they are in the United States. But those who cry "genocide" and "exploitation" in the use of overseas populations are not likely to assign exclusive blame to their fellow national researchers. For ethical and political reasons, they will assign heavy responsibility to the United States.

Fortunately, United States granting agencies have increasingly recognized these ethical problems. Since 1966 National Institutes of Health has required all overseas research grantees to give assurance of compliance with its requirements for peer review, though in at least one country where special cultural conditions seem to make peer review particularly hard to institutionalize, the N.I.H. accepts a suitable alternative form of ethical screening. As mentioned above, more recently, both the Population Council and the Ford Foundation, which are large suppliers of funds for overseas research in reproductive biology, have required from their overseas grantees peer review or some alternative that is still effective while being sensitive to the special cultural conditions of each country. Further, the Ford Foundation uses a special ethical consultant in addition to the ethical scrutiny that is provided by its scientific referees whenever human subjects are specified in grant proposals.[7]

### Ethical Problems Peculiar to Reproductive Biology

Beyond the general ethical problems that research in reproductive biology and contraception shares with all research on human subjects, it has some special ethical problems of its own. These problems arise from the special values, beliefs, and attitudes that human beings have toward their reproductive systems and behavior. These values, beliefs, and attitudes are different from those we have toward such other bodily systems and behavior as, for example, the digestive or the respiratory. For a long time, all our bodily systems have been enveloped in ignorance and error. But in addition to ignorance and error, sexual behavior has been controlled by taboo, sacred tradition, and religious belief. Until quite recently, sexual behavior has been approved of only for the explicit purpose of reproduction. Sexual pleasure for its own sake has been considered socially dangerous and religiously profane. Perhaps in the past, when reproduction itself and all life were so chancy, when the sheer goal of reproducing the population seemed always to be in doubt, the taboos and religious controls on sexual behavior for reproduction had their essential function. In modern times, with different values and with different fears—now more of overpopulation than of failure to reproduce the community and the species—people are more likely to approve of sexual behavior for the purpose of sexual pleasure without reproduction. Still, old ways die slowly and hard. There remains a minority, and often a very vigorous minority, that is adamant in its opposition to the new values, beliefs, and attitudes.

No wonder, then, that research in reproductive biology and contraception, which has been pursued intensively for only some forty or fifty years, has been continually surrounded by ethical protests and conflicts. At the beginning of this period, even in the countries of the industrial West, the large majority of the population viewed any "artificial" means of contraception with strong disapproval. As a teenager growing up in Cambridge, Massachusetts, during the 1930s, I happened by chance to know Gregory Pincus, one of the very first scientists to devote himself to the discovery of a biochemical method of controlling conception. I well remember the distaste and covert suspicion that even middle-class people who knew about his work directed toward him. Values, beliefs, and atttitudes have changed massively since then; we have experienced one of the fundamental social changes of our time. As all the polls of people's values and beliefs and all the surveys of their actual behavior show, the majority has now become the minority. Perhaps the change has been most striking and most unexpected among the Catholics; though official church doctrine against "artifical" birth control has persisted, and for a long time, but no longer, seemed to control the sexual behavior of the church's communicants. But a minority that feels it is unethical for people to use "artificial" birth control still persists and still holds unethical much of scientific research in reproductive biology. The objections of this minority have been expressed, first, against all "artificial" birth control, then against abortion as a backup method, and now against a variety of other research areas in reproductive biology, sometimes those that are connected with contraception but sometimes not—as for example, research using fetal tissue, the exploration of the possibilities of *in vitro* fertilization, and genetic screening.

Ethical beliefs and legal rules in a society do not stand in any neat one-to-one relationship. Nevertheless, there is some straining toward consistency between them. As the newer values and beliefs about contraception and the associated research in reproductive biology have developed, therefore, laws at both the state and the federal levels in the United States have tended to change to support the new majority. Legal action at the state level has been followed by Supreme Court decisions with regard, first, to sale and use of "artificial" birth control techniques and, second, to the use of abortion within prescribed time limits of the pregnancy period. Despite these rulings, the new minority still considers the behavior sanctioned by law to be unethical. Hence their continued protests and even efforts to overturn or circumvent the law. They are only doing what the old minority used to do when they were in the minority. In the 1930s, the minority in favor of "artificial" contraception engaged in protests against, and evasions of, the existing law.

During all this recent ethical conflict over contraception and research in reproductive biology, scientists—both as scientists interested in the advancement of research and as citizens with certain values with regard to sexual and social behavior—have tended to be counted among those standing for the new values and behavior. They have been supporters of, and sometimes leaders in, the great changes that we have seen in this area. Since they themselves have not infrequently not made it clear when they were speaking as scientists, when as citizens, or further, since they have sometimes declared themselves to be talking scientifically when they were only declaring their values as citizens, their opponents have sometimes put science itself under attack. This might happen anyway, but it is perhaps less likely to occur if scientists would more responsibly make it clear when they are stating scientific knowledge and when they are stating their preferred social values. Not that scientific knowledge does not have an important bearing on values; it most certainly does. Still, scientists should be careful to separate their knowledge from the values they are partly basing on that knowledge. Of course, they should be even more careful when they do not have knowledge at all or even reliable knowledge as a basis for their values. With the increasing public scrutiny and control over science that has come to exist as its influence on people's lives has so greatly increased, these cautions assume greater importance. Much of this public scrutiny is responsible, and scientists should respond in kind, with responsibility, both for its own sake and to forestall excessive and unwarranted controls on scientific research. Scientific responsibility earns and maintains for science the public trust, which provides the essential moral and material resources for continued scientific progress.

This public trust is essential as research in reproductive biology moves more heavily into new fields.[8] There is already much public discussion and some ethical protest over such procedures as amniocentesis for discovering the sex and other genetic characteristics, including defects of the fetus when it is still at a stage when it can safely be aborted. The same is true of the ongoing research efforts to achieve human fertilization *in vitro* with successful subsequent implantation in the human uterus. Fertilization has been achieved, but implantations have not yet in any important sense been successful. Fortunately, the research scientists most actively involved in this work have become increasingly concerned for their ethical responsibilties: public disclosure, informed consent from subjects, and careful consideration of the risk-benefit ratio.[9] Fulfillment of these ethical responsibilities is indispensable for maintaining public trust, in this and other new fields of research in reproductive biology.

## Conclusion

In conclusion, we may note again the fundamental fact that there is a gap or lag between the capacity of researchers in reproductive biology to carry on their scientific responsibilities and their capacity to live up to their ethical responsibilities. The gap or lag is slowly being closed, but there is still work to do. The power of biological research and the importance of its ethical consequences are not likely to be reduced in the foreseeable future. The dilemmas of science and ethics are going to be one of our continuing problems. These are not problems that can or should be solved by the scientists alone, though obviously they have an important part to play. Interdisciplinary cooperation will be necessary, from working scientists, lawyers, ethical philosophers, and social scientists. For the concrete and urgent ethical problems that scientific discoveries will pose, we need many different kinds of knowledge and perspective. One more thing is clear. We can no longer leave discussion of these matters to hearsay, allegation, impression, prejudice, self-interest, and ideology. Closer approximations to ethical responsibility will require more reliable and tested knowledge about just what the facts actually are in any ethical problem. Ethical dilemmas are often hard enough to resolve even when the relevant knowledge of the facts at issue is available. Without such knowledge, we are at a great loss. It is very important, therefore, that funds for ethical work and research on the part of those social scientists, ethical philosophers, lawyers, and working scientists who are specially interested in these matters be available. Such funds should be seen as having an essential function to play in the continued progress of research in reproductive biology and contraception.

# Notes

1. Much of the first part of this essay is based on published summary reviews and unpublished statements before the U.S. Congress on pending legislation in the area of human experimentation.
2. Thomas H. Marshall, *Citizenship and Social Class* (Cambridge, England: Cambridge University Press, 1950).
3. *New York Times*, January 9, 1973.
4. Henry K. Beecher, "Ethics and Clinical Research," *New England Journal of Medicine*, 274 (1966): 1354–60; M. M. Pappworth, *Human Guinea Pigs: Experimentation on Man* (London: Routledge & Kegan Paul, 1967).
5. These findings, and others reported later herein, are the result of investigations by my colleagues John J. Lally, Julia Loughlin Makarushka, Daniel Sullivan, and myself. They are reported in Barber et al., *Research on Human Subjects: Problems of Social Control in Medical Experimentation* (New York: Russell Sage Foundation, 1973).
6. Bradford Gray, *Human Subjects in Medical Experimentation* (New York: Wiley, 1975).
7. For an analytically organized and comprehensive collection of materials on the ethics of experimentation, see Jay Katz, ed., *Experimentation with Human Beings* (New York: Russell Sage Foundation, 1972).
8. A recent overview of ethical problems in medicine can be found in Jay Katz, *The Silent World of Doctor and Patient* (New York: Free Press, 1984).
9. See e.g., R. G. Edwards and David J. Sharpe, "Social Values and Research in Human Embryology," *Nature* 231 (1971): 87–91. For a layperson's critical review of this field, see A. Etzioni, *Genetic Fix* (New York: Macmillan, 1973), chap. 2.

# 12

# Research on Research on Human Subjects: Problems of Access to a Powerful Profession

As the sciences in general and the multifarious scientific specialties in particular seek to establish themselves in ongoing social systems, they face the same problems that all emergent social activities face. That is, they confront the problems of establishing identity, of winning legitimacy, and of securing those necessary social resources—such as recruits, money, and willing partners, or "access" as we shall call it here—without which they cannot function. Some essentials of the process of the secure establishment of identity, legitimacy, and resource bases for various general and special fields in the physical and biological sciences have been magisterially delineated by Joseph Ben-David (1971). The task of sketching this process for sociology in general or for particular sociological specialties requires much further scholarly attention than it has yet received, though some interesting initial attempts are now available.[1]

As a contribution to this task, we should like to analyze some experiences we have had in gaining willing partners, or "access," in some research we have done on research on human subjects in biomedical science (Barber, Lally, Makarushka, and Sullivan, 1973). As we defined our research, it was primarily a part of the sociology of science, though our actual analysis has also used concepts and findings from such other sociological specialties as the sociology of the professions, the sociology of medicine, and the sociology of deviance. The identity and legitimacy of the sociology of science and of these related specialties was of special concern to us, then, as we undertook our research.

Identity and legitimacy were all the more of special concern to us because we were very much aware of the very great prestige and power of the biomedical research profession. This is a group made up for the most part of university professors with both medical and scientific certification and achievements. They have both the organizational and the individual professional power to grant or deny to sociology-of-science researchers access to their views and behavior. They are not like the powerless poor whom sociologsits find it easy to study because the poor have neither the power nor the knowledge to raise questions about sociologists' identity and legitimacy or to deny access for these reasons.

Not only, then, were we very much aware of the power of the researchers using human subjects to deny us access, but also it was our strong impression that our identity as sociologists of science was very likely to be unknown and our legitimacy likely to be questionable. This impression was based on our own long experience in the sociology-of-science field and was supported by views such as those of the historian of science, Derek Price (1965):

> . . . the new knowledge about modern science seems to be growing in the midst of strong resistance from both within and without the field of the history of science. The resistance from outside is from the scientists themselves and is traditional. It has always been part of the special mystique of the scientist that he and he alone can really know about science. Only an esteemed and successful creative scientist can speak for his peers, criticize the state of science, counsel governments and universities, and guide the policies of laboratories and learned societies. An outsider must be presumed ignorant, not merely of the technical facts, but also of that special knowledge of the life of science that can only be won on its battlefields.

As a result of our anticipation that many biomedical scientists would share the views outlined by Price, we felt that just presenting ourselves as sociologists of science would not be enough to assure us of the access we desired. We, therefore, used three other techniques in order to gain the necessary legitimacy. First, we called upon other roles, goals, and values we shared with the researchers we wanted to study. Second, we suggested to the biomedical reseachers that we had access to some possible countervailing power. Finally, we took advantage of the conflict of values biomedical researchers sometimes feel because their norms and obligations as individual professionals can conflict with those which go with being members of research organizations. We did not always quite know just what we were doing, nor was our eventual effectiveness in gaining access completely planned; we "lucked into" some of our success. In retrospect, however, we can see fairly well just what techniques we used and why they led us to our goal. And because they are of general significance in doing research on powerful professions, we present them here for other interested social researchers.

Before proceeding to the details of our analysis, a very brief account of the purposes and procedures of our research will be desirable. We wanted to discover the prevalent patterns of expressed standards and reported behavior among biomedical researchers using human subjects with regard to two key ethical issues, informed consent and the ways in which risks of physical harm to subjects are counterbalanced or not by therapeutic benefits for the subjects, what we have called the risks-benefits ratio. We were interested in these two issues because of their importance in all the relevant ethical codes, in the Department of Health, Education and Welfare's *The Institutional Guide to DHEW Policy on the Protection of Human Subjects*, and in statements made to us by physician-researchers during our preliminary field work. Our goal was to discover some of the determinants of "strict" and "permissive" ethical standards and practices among biomedical researchers and to make an estimate of the degree to which human subjects are in any sense "misused" in biomedical research. We looked for determinants of "strict" and "permissive" standards and practices in a number of places: in what we called "the dilemma of science and therapy"; in professional socialization patterns; in the structure and processes of peer-group review; and in various informal small-group interaction structures and processes. We were, then, asking our subjects for more than the mere time involved in responding. We were asking for information about controversial and sensitive matters which might provide the basis for evaluating the "ethicality" of them and their institutions.

To get at these patterns and their determinants, we did two different studies. One, our National Survey, was a mailed questionnaire study completed by representatives (usually the research directors) of a nationally representative sample (292) of biomedical research institutions using human subjects. Our second study, Intensive Two-Institution Study we called it, was based on one-hour or longer personal interviews with about 350 research physicians who actually use human subjects. These researchers were located in two different organizations: most of them in University Hospital and Research Center; a minority of them in Community and Teaching Hospital. In our National Survey we concentrated on patterns of expressed standards about the consent and risk-benefit ratio issues and on the structures, processes, and efficacy of peer review groups. In our Intensive Two-Institution Study we were again interested in these same matters and in additional matters as well. In this study, for example, we collected self-reports of the researchers' own behavior with regard to the risk-benefit ratio issue and we studied patterns of professional socialization and of small-group interaction processes.

To repeat, knowing that it was very likely that we had neither identity nor legitimacy as sociologists of science or medicine, we saw or came upon three other ways to gain access.[2] Let us look at each of these ways in some detail.

## 1. Shared Roles, Goals, and Values

Since humane therapy is one of the central values of the medical profession, in its research as well as in its therapeutic branches, we knew we could appeal to this shared value in justifying our interest in the patterns and problems of human experimentation, where the humaneness of existing procedures has very much been called into question by professionals and laity alike.[3] Accordingly, our letter to potential respondents in our first study, the National Survey, opened with the following statement:

> As you are probably well aware, the whole area of human experimentation has come under vastly increased discussion in the last couple of years. There has been, however, no systematic empirical research attempted which aims at discovering facts that would better inform the whole discussion. The enclosed questionnaire is the first part of some extensive research designed to remedy that situation.

And we closed with the following statement, which repeats our statement about the importance of this common goal of learning more about how to achieve humane therapy in human experimentation:

> In view of the vast importance of the topic of human experimentation, the great paucity of information about it, and the potential use to you of the data we are trying to collect, we would greatly appreciate it if you could complete the questionnaire. If you wish, we shall be happy to send you a brief statistical summary of the findings of this survey. There is a space on the questionnaire for you to indicate whether you would like such a summary.

In our second study, Intensive Two-Institution Study, of course, we sounded this same theme. In the first sentence of our letter requesting a personal interview, we spoke of "the difficult, complex, but very important problem of the use of human subjects in biomedical research." We expected that this shared goal would help us to gain access and it probably did. In both of our studies we were given much positive evidence that the goal of humane therapy and the problems of establishing proper procedures in human experiments were of considerable importance to our respondents. For example, in the course of their interviews many respondents were eager to know what we had learned so far, how "most" researchers had answered, how typical their own responses were. We were not only requesting cooperation, therefore; we were offering something in return, i.e., information about the solution of ethical dilemmas faced by our respondents.

It will have been noticed that we have already touched upon another shared value, the need for science and systematic research to help in the solution of human problems. It was a deliberate appeal on our part to this shared value that made us speak of "systematic empirical research" and "extensive research" in our very first paragraph of the letter asking for cooperation in the National Survey. And in our letter for the Intensive Two-Institution Study, when we had already completed our first study, we spoke of ourselves as what we felt we had proudly become, a "research group." Although there is and continues to be much resistance to social science from the physical and biological scientists in our society, still there is almost always some ambivalence in their stance. How can they, who hold the value of science so strongly, not grant at least some aid to those who claim to share that value? Probably the same holds true for laypersons: there must always be some ambivalence, some positive value, in their negative view of social science in general or of some particular social science specialty or individual research project. It is under the covering general value of science, now so much more solidly established than it was in even the seventeenth century, that we social scientists have, for all our difficulties, an easier road to travel. Because of this shared value, social scientists have what the economists refer to, in another context, as "the advantages of the latecomer."

Since nearly all of the respondents in the Intensive Two-Institution Study and very many in the National Survey were professorial or other members of universities, we were very careful to point out that we shared this role with them. Our letterhead clearly indicated our university membership and rank, and our university affiliation was also printed on our mailed questionnaire. In the text of the letter for the National Survey we even said explicitly, "the research is being carried out at Columbia University." University membership is a warranty of standards of scientific and ethical performance that are very important to the biomedical-research professionals to whom we needed access. The acceptance of sociology into the American university, to be sure often initially for practical reasons (e.g., rural sociology and social work) rather than for fundamental-science reasons, has been a fact of the greatest importance in the development of American sociology. The flexibility and openness of the American university, whatever their disadvantages, have had the function of providing a general identity and legitimacy for sociology and the other social sciences that has been indispensable in their development. It was that general identity and legitimacy to which we were laying claim with our university role-partners in the biomedical-research profession.

We had one other claim to legitimacy to make. We knew that the American private philanthropic foundation has probably nowhere been of more consequence and is probably nowhere held in higher esteem than in the university biomedical-research profession. Like the university, the foundation is a warranty of standards of scientific and ethical performance that are important to the researchers whom we needed as willing partners. In our first letter, therefore, we mentioned that our work was being carried out under "a grant" from the Russell Sage Foundation and in the second letter we spoke of "continuing grants." The Russell Sage Foundation was also identified on our mailed questionnaire. Although the Russell Sage Foundation has never made grants for biomedical research, we felt sure that its generic status as a foundation would be an important legitimizing agent for us.

## 2. Countervailing Power

There is no doubt, despite these several value, goal, and role sources of legitimacy, that biomedical research professionals using human subjects in experiments still had the power to deny our requests for access. Indeed, many institutions in the National Survey and many individuals in the Intensive Two-Institution Study did deny us, though apparently not in sufficient numbers to destroy the representativeness of the samples we were studying. We had, however, what *may have been* (we cannot test this suppostion) one source of countervailing power, the funding and regulatory power of the National Institutes of Health (N.I.H., Public Health Service). If this power did operate, it was certainly unintended by the N.I.H., and we ourselves sought to make it clear that we did not want it to operate.

We needed the help of the Division of Research Grants of the National Institutes of Health because we felt that data in its possession were the best source for a nearly total population or at least for a representative national sample of institutions for our National Survey. The Division of Research Grants, we learned, had a computerized list of the names and addresses of the responsible officials of all the biomedical-research institutions that had filed assurances of compliance with the N.I.H. regulation, Protection of the Individual as a Research Subject, which has more recently been transformed into the D.H.E.W. *Institutional Guide* referred to above. Because the N.I.H. was providing about 35 percent of all the funds available for biomedical research in this country, and because we felt that, therefore, no research institution could afford not to apply for these funds and to file the required assurance, we felt that the N.I.H. was probably the only complete and available list of institutions doing research on human subjects. Upon request, the Division of Research Grants was kind enough to furnish us a computer printout of this list, and we had the beginnings of our sample.

In writing directly to the responsible official of each institution on the N.I.H. list, we felt obliged to report how we had located him:

> The questionnaire you have received has been sent to all institutions in the United States that have complied with the request of the National Institutes of Health that an institutional review procedure be set up to review proposed research on humans. Your name was given to us by NIH as the person in your institution with whom they corresponded on this matter. The questions in the questionnaire are designed under the assumption that you will be able to serve as an informant on the matters pertaining to your institution which appears there.

We felt obliged to make this report, but we also felt obliged not to claim the power of the N.I.H. First, in the interest of the Division of Research Grants, we did not wish even to imply the false claim that N.I.H. required each institution to respond. The N.I.H. has been at great pains in this whole area of the regulation of the use of human subjects to deal cooperatively and not merely powerfully with the biomedical-research profession. We did not wish to misrepresent them in this instance. But, second, in our own interest, in the interest of asserting our own scientific autonomy, we did not wish to claim or be identified with the power of the N.I.H. For both these reasons, therefore, we went on to say in our letter:

> We would point out that while NIH has aided us in this research in certain small ways, such as the above, we are completely independent from them. Please be assured that your replies will be held in strictest confidence. No one besides the immediate Columbia researchers will be able to identify your questionnaire, and neither your identity nor that of your institution will be revealed. (The use of a number, rather than your name, at the bottom of the back page of the questionnaire helps to ensure this.)

Thus we tried to remove any possibility of the countervailing power of the N.I.H., with its funding and regulatory powers, coming into play against the power of the biomedical-research profession to deny us access. Since many institutions in fact did not respond, we seem to have been at least partially successful. And all the respondent institutions may have answered for quite other reasons than fear of the powers of N.I.H. Still, countervailing power may have come into play here. We would have liked to avoid all reference to the N.I.H. and not even raise this possibility of countervailing power. But we could not avoid such reference because of our felt obligation to report the source of our list of institutions and names. There may be, of course, some who feel that the countervailing power of the N.I.H. should have been used deliberately, to *command* access for us. Indeed, there are some who may feel that the N.I.H. should be using its power directly, to do studies like this itself. It seems likely, however, that here as elsewhere power has its definite limits and that claims to legitimacy are effective without the addition of more than a minimum of countervailing power. Sociology may well have to use some countervailing power as it seeks to establish its identity, legitimacy, and claims to social resources more securely, but surely both the morality and the effectiveness of sociology will be greater when it is thus securely established.

### 3. Individual Professionals versus Organizational Members

Biomedical research is now very much a collective rather than an individual enterprise. Researchers typically work in collaboration groups (our data showed that four out of five studies in our sample of 424 studies are carried out by collaboration groups of two or more individuals), and these collaboration groups are located in larger oranizations such as our University Hospital and Research Center and our Community and Teaching Hospital. So important are these collective contexts of research in providing identity, legitimacy, and resources to biomedical researchers, that our first thought in designing the Intensive Two-Institution Study, where we wanted personal interviews, was to ask permission from the two organizations that we had selected as representative after doing a cluster analysis on our sample of institutions in the National Survey. But the difficulties we might have by approaching the organizations directly occurred to us immediately. First, we feared the delay and the uncertain timing that would result if we had to wait while a decision to grant us permission was proceeding through channels and up the organizational hierarchy. By then, we ourselves were a continuing collaboration group and could not afford to wait to gain access. Second, though, we were concerned that reasons of organizational fearfulness or convenience would be raised to deny us access altogether. As we put it, somewhere along the way some executive, some lawyer, some trustee would counsel against giving us access. We were afraid that one strong-feeling individual would carry the day against his at best neutral colleagues. After all, we could hear these colleagues say, "Who are these fellows? What do they know about these esoteric matters? What right do they have to push their way in here?" And if such colleagues had any of the ambivalence we have spoken of above, hesitating a little to deny the claims of "scientists," even if they were only self-styled, we could hear them say, "We're not against their study, but let them do it elsewhere."

For these reasons, we decided to approach our biomedical researchers directly, as individual professionals, rather than as what their roles also made them, organizational members. But we still had to give the organizations their due. We had meetings with the responsible high officials of the two selected organizations, made our claims for identity and legitimacy to them, and indicated that we intended to approach each researcher as an individual, leaving to him the right to grant or deny us an interview. So strong and so taken for granted is the norm of individual professional autonomy in these organizations (the responsible officials were themselves men who had done research) that assent was readily and gladly given.

Further evidence of the strength of the norms supporting the autonomy of the individual professional role came from the individual researchers themselves. Not a single one of the individuals who gave us an explicit reason for refusing an interview mentioned our lack of organizational legitimation. And among the approximately 350 individual researchers who did give us personal interviews, only one ever raised the question of organizational permission. Despite the great importance of their organizational membership, these biomedical researchers took it very much for granted that they had the right to speak freely about the ethical aspects of their research activities. The professional's assumption of the right to autonomy in speaking about some of the most important activities he carries on in his research organization has no counterpart in any rights that could be claimed by individual members of business organizations.

Thus individual professional autonomy, even when professionals are also organizational members, can be of considerable help to social researchers seeking to gain access to the activities of these professionals and their organizations. We do not mean to say, of course, that individual professional autonomy either does not or should not have limits set by the organizations in which they are members. Indeed, in the area which we were studying, the ethics of research on human subjects, a central focus of our attention was on the way in which older patterns of individual professional autonomy were being controlled by such organizational devices as peer review groups, to bring them into better accord with the standards of socially responsible professional behavior. We are not interested here in what the relationship between individual professional autonomy and organizational membership has to be to achieve different specific social outcomes. Our interest here is only methodological, to point out that when a social researcher is seeking access to powerful professionals of any kind—medical researcher, lawyer, or academic—he very much needs to know just what the relationship between individual professional autonomy and organizational membership is in the class of situations or specific situation he is studying.

*In summary*, we have tried to explicate three ways in which some sociologists of science gained access to members of a powerful profession, ways we think have generic significance for those doing research in such a context. It will perhaps have been noticed that we have treated the research situation very much as we would have treated any other situation of social interaction. For that is very much what the social research situation is. As we attempt to synthesize and generalize our knowledge about social research as a social-interaction situation—not only to have the knowledge for its own sake but to provide us and our students with a methodological canon—we need to bring to bear the concepts that we use everywhere in social life: power, legitimacy, roles, professions, norms, values, and the like. By understanding better the social-research situation as interaction, perhaps we can even come to understand interaction in general somewhat better.

Finally, it will be useful to make explicit one of the several implications our experience with the biomedical research profession has for us professional sociologists ourselves. The research experience we have just reported, and, of course, the research experience of many of our sociologist colleagues, shows that we sociologists are not without considerable power to gain access to social situations where deep values and interest are involved. By our research in such situations, we may have the power to do considerable good or considerable harm. Such power, though probably less, is not unlike the power that the biomedical research profession has in its area, or that other professions have in theirs. In short, we have to be aware that we can no longer consider ourselves powerless or be unaware that, whatever our own view of ourselves, others may consider us powerful.

This being the case, we have to face at least two ethical problems if we are to be a socially responsible powerful profession.[4] The first is that we must not put our own needs and rights as scientists above the needs and rights of those whom we study, whether they are powerful or powerless. Like the biomedical-research profession, for example, we must be respectful of the civil right of those we study to be asked their informed consent to be so studied. An essential element of such consent is full information about the probable risk-benefit ratio for those being studied.

It is our impression that professional sociologists have not been overly zealous in this first of our ethical responsibilities. While the American Sociological Association has made fitful and partial efforts toward institutionalizing ethical codes and procedures, there remains what may be considered, at best, widespread ambivalence toward them among the researching members of the profession and, at worst, considerable indifference and hostility. Certainly, professional sociologists have not been innovative and intensively energetic in this ethical concern in a way of which we could be rightly proud. The professional psychologists, with their codes of ethics (now finally in its third edition), have done much better.

The second of our ethical problems as a socially responsible profession is to make ourselves accessible, tolerant, and responsive to relatively objective scrutiny by outsiders. Since professional sociology, like all professions that have social and political consequences, at some point becomes too important to be left to the professionals themselves, we must expect such scrutiny from journalists, "humanists," other social scientists, ideologists and social critics, and "the community." Again, it is our impression that sociologists are at best not much more responsive to such "criticism" than many of the other professions that they define as much more powerful than themselves. We may think that internal criticism, of which there is a good deal, is enough, but it is not. We cannot responsibly shut ourselves off from outsiders whom we affect, from their values and interests, for which only they themselves can often best speak.

In sum, our analyses often apply to ourselves. As we define the problems of access to a powerful profession, we must be careful to consider what this means for us as one of the powerful professions.

## Notes

1. On sociology in general, see Coser (1971); on French Sociology, see Terry Clark *Prophets and Patrons: The French University and the Emergence of the Social Sciences* (Cambridge: Harvard University Press, 1973); and on medical sociology, see Barber (1968).
2. For another discussion of access, see Form (1971).
3. For criticisms from within the profession, see Beecher (1970) and Pappworth (1966). For a mixture of professional and lay discussions, see Freund, ed. (1969) and Katz, ed. (1972). The lay press has recently been filled with reports and comments on the ethics of the Tuskegee syphilis case.
4. For a generalized discussion of the social responsibilities of powerful professions, see Barber, Lally, Makarushka, and Sullivan (1973), chap. 10.

## References

Barber, Bernard 1968 "The Functions and Dysfunctions of 'Fashion' in Science: A Case for the Study of Social Change." *Mens en Maatschappij* 43, no. 6 (November–December): 501–14.

Barber, Bernard, John J. Lally, Julia Loughlin Makarushka, and Daniel Sullivan 1973 *Research on Human Subjects: Problems of Social Control in Medical Experimentation*. New York: Russell Sage Foundation.

Beecher, Henry K. 1970 *Research and the Individual: Human Studies*. Boston: Little, Brown.

Ben-David, Joseph 1971 *The Scientist's Role in Society: A Comparative Study*. Englewood Cliffs, N.J.: Prentice-Hall.

Clark, Terry N. 1973 *Prophets and Patrons: The French University and the Emergence of the Social Sciences*. Cambridge: Harvard University Press.

Coser, Lewis A. 1971 *Masters of Sociological Thought*. New York: Harcourt, Brace, Jovanovich.

Department of Health, Education, and Welfare 1971 *The Institutional Guide to DHEW Policy on Protection of Human Subjects*. Washington, D.C.: U.S. Government Printing Office.

Form, William H. 1971 "The Sociology of Social Research," in R. T. O'Toole, ed., *The Organization, Management, and Tactics of Social Research*. Cambridge, Mass.: Schenkman.

Freund, Paul A. ed. 1969 "Ethical Aspects of Experimentation with Human Subjects." *Daedalus*, Spring.

Katz, Jay, ed. 1972 *Experimentation with Human Beings*. New York: Russell Sage Foundation.

Pappworth, M. H. 1966 *Human Guinea Pigs: Experimentation on Man*. London: Routledge & Kegan Paul.

Price, Derek J. De Solla 1965 "The Science of Science," in John R. Platt, ed., *New Views of the Nature of Man*. Chicago: University of Chicago Press.

# 13

# Liberalism Stops at the Laboratory Door

As the scope, power, and funding of both biomedical and social-science research have increased hugely in recent years, so also, inevitably, has the use of human subjects. Ever larger numbers of experimental and other research subjects have been put at risk, physical risk in biomedical science, social and psychological risk in the social sciences. This increased risk factor alone, on quite rational and utilitarian grounds, would have called for some new measures of social control over the use of human subjects in research. But there is an additional factor, a value factor, crying out for more ethical controls on the use of experimental subjects. That factor is the increased emphasis everywhere in the world, and among all social groups, on the value of equality. Women, blacks, youth, children, the poor, students, people in the less developed countries, patients, and subjects of research—all these feel themselves treated unequally, exploited by the more powerful. They want more full and effective participation in the vital decisions that determine their welfare. Even benevolent parternalism is abhorrent to them.

So we have a new social problem, exposed more often by public scandals, such as the Nazi doctors' use of Jewish subjects; thalidomide; the Southam-Mandel injection of live cancer cells into geriatric patients without their informed consent; the Tuskegee scandals concerning syphilis experiments; Stanley Milgram's deliberately deceptive psychological experiments on obedience using human subjects; and Camelot, the deceptive social science studies carried out in South America on which Irving Horowitz, among others, blew the whistle, than by rational inquiry or by systematic ethical self-monitoring by the research professions. But this is not another one of those social problems that we social scientists can just study. It is our own social problem and we have to take some action on it.

How have the biomedical- and social-science-research professions responded to this new social problem? On the whole, with indifference and hostility. I have heard distinguished medical researchers say that new government regulations on the use of human subjects mean "the death of science." And I have heard both medical and social researchers speak angrily about these new government regulations in the same tone and with the same arguments that businessmen in the 1930s spoke of such new government regulations as the Securities and Exchange Commission. But most of the researchers I know are more indifferent than hostile. Not so disturbed as the angry men, they still share the latter's emphasis on the absolute autonomy of researchers. Science is what excites and rewards them; it is their primary value; and for the time being their ethical responsibilities for the consequences of research to the subjects of research are of no great concern to them. They want the status quo preserved; they reject innovations in the area of social control either from their fellow researchers or from "outsiders" like the government. In sum, they are powerful and conservative. Only government action, beginning with N.I.H. regulations in 1966 and continuing through further action by D.H.E.W. and now by the National Commission for the Protection of Human Subjects, has moved them to improve their ethical standards and practices.

The conservatism of biomedical and social researchers is all the more interesting in the light of the fact that these groups are otherwise liberal. That is, they are generally for social reform on behalf of the underdogs, or those who feel themselves unequal. Biomedical researchers, for example, have often been in favor of changing systems of payment for medical care in the direction of more equality. And social scientists, as systematic survey data from Lazarsfeld, Stouffer, and Lipset and Ladd have shown, are liberal on nearly all social issues. It would seem that liberalism stops at the laboratory door. Scientists have so exclusively devoted themselves to the goals and values of scientific research that they turn conservative when those goals and values have to be accommodated to other important social goals and values.

Unfortunately, space allows me to present only some of the evidence we have to support what I have said about the conservatism of researchers with regard to the ethical problems of research on human subjects. The evidence we have for the lack of ethical concern among biomedical researchers is based on systematic studies; that on social researchers is more impressionistic, but still weighty.

I should establish first, of course, that the existence of ethical delinquencies among biomedical and social researchers is not just a matter of scandalous allegation. Two studies carried out by my Research Group on Human Experimentation—one on a nationally representative sample of biomedical research institutions, the other on 350 researchers using human subjects at University

Hospital and Research Center—and research done by Dr. Bradford Gray at another university medical center have resulted in the following findings: a definite minority of researchers are "permissive" in their standards on both the informed-consent and risk-benefit ratio issues; a definite minority of researchers carry out investigations in which, according to their own estimates, the risks exceed the benefits; the poor who are ward and clinic patients are more often the subjects of research than private patients; moreover, these ward and clinic patients are more often the subjects in the unfavorable studies where risk exceeds benefit to the subject and sometimes even to future subjects; and, finally, about one-third of the subjects on a study done in a relatively high-standard university medical center did not know they were experimental subjects until so informed by a social researcher. It is not likely that the situation is notably better in social research.

The priority of science over ethical concerns can be seen in our data on medical schools, where presumably the central values of the research profession reside and are taught to new recruits. The evidence from our interviews with 350 researchers shows that not much training in research ethics is given in medical schools. Only 13 percent of these researchers reported that in medical school they had had a seminar, a lecture, or part of a course devoted to the ethical issues involved in the use of human subjects. Only one researcher said he had a complete course. Thirteen percent of the respondents said that the ethical issues of research had come up when, as students, they did practice procedures on one another; 24 percent said that they became aware of ethical issues in doing experimental work with animals; 34 percent remembered discussions with instructors or other students of the ethical issues involved in specific research projects read about or discussed in class. However, and this is the significant finding, 57 percent of the investigators we interviewed reported not a single one of these learning experiences in research ethics.

Further evidence, from the National Survey, shows that medical schools have not been ethical leaders either with regard to peer review, one of the chief ethical monitoring mechanisms now widely used as a result of the 1966 N.I.H. mandate. The medical schools' record on peer review has been less good than that of other types of biomedical-research institutions. They were less likely than these other types of institutions to have set up a peer-review group before N.I.H. required one, less likely to have had one that met N.I.H. guidelines in 1966, less likely since then to review all clinical research, less likely to have included nonmedical and medical outsiders on their peer-review groups as recommended by N.I.H., and less likely to have had their review groups well received by their individual researchers.

Medical schools, the Association of American Medical Colleges, and professional associations of clinical researchers have been very much quicker to lobby for research funds and to protest funding cuts in Washington than they have been to organize seriously for studying and making policy on the ethics of research. Their initiatives and efforts in ethics hardly match their initiatives in science.

Finally, we can see this same emphasis on science as against ethics among individual biomedical researchers. Our data show that ethical concern for research subjects is not a highly salient consideration for working researchers when they select their collaborators; scientific ability is. When we asked our 350 researchers, "What three characteristics do you most want to know about another researcher before entering into a collaborative relationship with him?" 86 percent said "scientific ability," 45 percent said "motivation to hard work," 43 percent said "personality," and only 6 percent said anything that means "ethical concern for research subjects." Saliency and actual concern are not the same thing, of course, but it is clear, nevertheless, that science is more important than ethics as a criterion for choice of biomedical research collaborators.

Now let us look quickly at some unsystematic evidence on the indifference and hostility of social scientists to concern with research ethics. The lack of systematic evidence is perhaps itself significant. Sociologists, who sometimes seem to study everything, have not systematically studied the training that social scientists give their graduate-student recruits in the ethics of using human subjects. Of course, any such study would probably find that as little of this training occurs in medical schools as in social-science graduate schools.

Looking at the psychologists, we can see very clearly the resistance and hostility that active researchers feel toward ethical self-regulation in the use of human subjects. It was not until the third edition of its ethical handbook that the American Psychological Association included a section on the ethics of research, and this came after the N.I.H. mandate of 1966. There has been a great deal of resistance among psychologists to an effective statement on the ethics of using human subjects, much of it expressed quite openly in the pages of the *American Psychologist* during the last few years. As a member of the consulting committee for the new edition's section on research ethics, I saw some of the preliminary drafts. In a near-final one, the whole matter of the ethics of using human subjects was referred to as "public relations."

Have we done better in sociology? I think not. I have lost count of the number of times I have been on the American Sociologists Association (ASA) Ethics Committee, but it is at least three times. Only once was the committee effective enough to spend a couple of years of hard work, under the chairmanship of Robert Angell, drawing up a detailed code of ethics. As late as the late 1960s, when we presented that code to our council, it was politely rejected by being tabled. As a member of the present committee,[1] I have the same sense of indifference and hostility on the part of the powers-that-be in the association. My sentiments are shared by our esteemed colleague, Gideon Sjoberg, the present Ethics Committee chairman, who has a long record of concern with professional ethics. In a letter to the ASA. Council, Sjoberg wrote: "In my judgment, the Committee on Professional Ethics serves, as it presently functions, largely as 'window dressing' for the Association. In order to rectify this situation, the Council will need to deal with jurisdictional issues, to work toward a reformulation of the Code and its procedures, and to provide adequate funding for the Committee." I think that the reason the council does not take the Ethics Committee seriously is because council members, like most social-science and medical researchers everywhere, put their science goals and values ahead of the ethical problems resulting from science. When a researcher member of the ASA complained to the council and the ASA Washington staff that the new D.H.E.W. regulations on the ethics of the use of human subjects were, as he put it, "impossible" to live with, the matter was referred to the Committee on Academic Freedom and not to the Ethics Committee. We of the ASA are obviously more concerned with our own scientific freedom than with our ethical responsibilities to the subjects of our research.

A few last words. The research professions are powerful in American society because of their control over powerful and esoteric knowledge. They make large claims to the necessity for a considerable measure of self-regulation, claims that are justified because of this esoteric knowledge, though the power that this knowledge gives also mandates an equally considerable measure of public involvement in and control over professional activities. Our goal must be an effective mixture of professional self-regulation and public accountability and responsibility. At the present time we are so beguiled by our pursuit of science that we have devised neither effective self-regulation nor adequate public responsibility. Surely it is not beyond the wit and virtue of social scientists and other researchers using human subjects to engage in a little self-reform, a little constructed social change, that will bring us somewhat closer to our claims to professionalism.

## Note

1. Though I realize that this section is out of date, I have let it stand as a vivid reminder of how things really were as recently as the mid-1970s.

# 14

## Control and Responsibility in the Powerful Professions

Everywhere in the United States the professions have reached new heights of social power and prestige. Everywhere, because of the power of their special knowledge, they are of increasing consequence in the lives of individuals and in the affairs of groups, the polity, and the society as a whole. Yet everywhere they are also in trouble, criticized for their selfishness, their public irresponsibility, their lack of effective self-control, and for their resistance to requests for more lay participation in the vital decisions professionals make affecting laypersons.

The signs of this trouble are manifest in many quarters. In California the governor has for the first time appointed laypersons to every one of the state boards that regulate professional conduct; formerly these boards were monopolized and dominated, run in their own interests, by the professionals themselves. In New York State the legislature has passed new laws requiring more effective peer control over delinquent medical practitioners.[1] Also in New York, the Board of Regents, which has responsibility for the public control of some twenty-nine "professions," has voted to permit professions to do limited advertising in newspapers and magazines. Further, it has issued a new regulation requiring health professionals to show a patient his medical records upon request, except when this "would adversely affect the patient's health."[2] This latter exception is an application of the therapeutic privilege, which sets certain limits on the patient's right to know, the doctor's duty to tell. In 1973 the U.S. Congress established the National Commission for the Protection of the Subjects of Biomedical and Behavioral Research to reduce long-standing abuses against the persons and civil rights of the subjects of biomedical and social research. In 1974 a report on medical malpractice was issued by a commission appointed by the secretary of the Department of

Health, Education, and Welfare.[3] As the problem of medical malpractice continued to become more severe, the State of New York appointed its own investigatory panel, which issued its report in 1976.[4] In the same year, another Senate committee, the Commitee on Government Operations, investigated the malfeasance of another powerful profession, the accounting profession.[5] The "arrogance" and elitism" of the academic profession, which trains and sets standards of self-control and responsibility for the other professions, have also recently been criticized.[6] So widespread and recurrent is public protest against the irresponsible power of the professions that two sociologists have referred to "the revolt of the client" as a now endemic phenomenon in American society.[7]

Obviously there now exists what sociologists call a "social problem," that is, a set of social conditions defined by at least some articulate groups in the population as not only morally bad but urgently in need of reform. The excessive power, the lack of satisfactory self-regulation, and the public irresponsibility of the professions is one of our new "social problems."

Just as obviously, at least to sociologists, there is a sociological problem here, or better, a set of them. If public policy is to be effective in achieving some reform of the present defects of the professions, it needs satisfactory analyses of several sociological problems that underlie these defects. What is "a profession" and how are control and responsibility essential characteristics of professions? What evidence or measures can we produce to describe the alleged defects of control and responsibility in the professions? And what reforms of the mechanisms of social control and public responsibility of the professions are now occurring or can be suggested on the basis of sociological analysis and data?

## A Definition of Professional Behavior

Professional behavior is defined in terms of three essential and somewhat independent variables: powerful knowledge, self-control or autonomy, and public responsibility or direct service to the public and the public welfare.[8] An occupation is the more professional the more it actually displays, not just claims to possess, of these three characteristics. What holds as between occupations, also holds within occupations. Not only are clergymen more professional by our sociological definition than morticians, but some clergymen, by the same standards, are more professional than other clergymen. There is always some range of professionalism within any large occupational category that is publicly defined as professional. But this range may vary in different categories of professionals. For example, the range between the most and least professional physician is less than between the most and least professional clergyman. The matter of degree of professionalism, whether between or within occupational categories is always important; the sociology of the professions needs to create better measures of these differentials than we now have. Such measures are needed for each of the three essential defining variables. A professional category may have more of knowledge and less of effective self-control and actual public responsibility. And its degree of knowledge, or control, or responsibility may be changing over time, to more or less. Such changes are matters of sociological and practical consequence. Now let us consider each of the three defining variables more closely.

*Powerful Knowledge*

There are two generalized bases of power or consequentiality in systems of human action.[9] One is knowledge, the other is the capacity for making decisions in the informal and formal organizations into which human action gets structured. By knowledge we mean to refer to the whole range of symbols or idea-systems which define the means and ends, the interests and values, the beauties and the ultimate meanings of human action. Though some occupations combine a mix of knowledge and decision-making, most tend to be characterized chiefly by one or another of these two generalized dimensions of power. The professions are those occupations that specialize in the development and application of powerful knowledge. Lawyers, for instance, specialize in the knowledge that affects our central interests, our deepest values, our very sense of equity and justice. Physicians control the knowledge that helps us achieve our interrelated senses of physical, psychological, and moral well-being. The theology and preachings of clergymen define our approach to good and evil in human affairs, our aspirations for meaningfulness, our ability to cope with human finitude and death. Academic scientists—physical, biological, and social—also define our frames of meaning by their theories of the nature of the world and its evolution, their understanding of the essence of life and its processes, and by their ideas about power and justice in social life. Artists of various kinds mold our sense and appreciation of beauty and meaning in the world.

Knowledge is always power, and each of the professions has its own special powerful knowledge. The more powerful professions, or parts of professions, are those whose knowledge is generalized and systematic yet with some application to the empirical physical, biological, or social worlds. The irony of power in knowledge is that the more it is based on generalized and systematic theory, the more, soon or late, it has important consequences for systems of human action. Common sense, with its tendency to stress the immediacy and specificity of power, rejects this paradox in the power of knowledge, but sociological analysis and research confirm it. The "semi-professions,"[10] or the semiprofessional parts of those professions where there is a range of control over generalized and systematic knowledge, are those with the more concrete, less organized knowledge.

*Effective Self-control or Autonomy*

Because generalized and systematic knowledge is esoteric, known to and controlled by only a relatively few occupational specialists as a result of considerable and continuing training and work, its development and application require a considerable amount of self-control or autonomy for those who specialize in these tasks. This self-control is effected through a number of different standard social control mechanisms, such as the inculcation of high cognitive and moral standards, the use of informal peer controls, formal organizational mechanisms within the profession, and, finally, the domination by the profession of a variety of external legal and political mechanisms in the area of licensing, standard-setting, and sanctions for deviant behavior. Autonomy and self-regulation in these various forms are not something illegitimately seized by the professions but approved by public opinion and granted by formal action of various political bodies. (Remember, we are presenting a definition here, not describing any actual social reality. That is a matter always requiring investigation.)

*Public Responsibility or Direct Service to the Public Welfare*

Just because of its consequentiality for human affairs, power of any kind is always restricted in some measure by social and political controls. The power of every kind of specialist is always too important to the public welfare for it to be left to the specialists themselves. In the case of professional power, the necessary granting of a considerable degree of autonomy to the professions imposes on them an equally considerable degree of obligation to exercise their power clearly and directly for the welfare of their clients and the public. Professionals acknowledge and are even proud of this special obligation. They value their special role as "fiduciaries" and "public servants." In the name of this special obligation to display what Talcott Parsons has called "collectivity-orientation," the professions ask for public "trust" and compliance with their "orders." A good deal of this trust is willingly given, but nonetheless, even for the professions, autonomy is never granted as an absolute. Both the public and its political representatives always exercise some controls to ensure the actual public responsibility and service behavior of the professions.

### Defects in Control and Responsibility in the Powerful Professions

Our analytical definition states the essential theoretical nature of the professions and the conditions under which professions operate for the best interests and values both of professionals themselves and of the publics they serve. Our definition provides *a measure* for actual empirical social reality, *a guide* for investigating the degree to which actual professions do or do not fulfill the essential criteria of professionalism.

It is probably the case that these essential defining characteristics of a profession have seldom been anywhere near fully realized in the history of actual professions. Berlant, for example, examining the history of codes of medical ethics, has argued that these codes have, from Hippocratic times right down to the present, contributed more to a monopoly of power and interest for the profession than was good for the public welfare.[11] Certainly no actual profession comes up to these standards at present. Certainly also, as we have seen, a variety of voices is now expressing its dissatisfactions with present defects in control and responsibility patterns in all the powerful professions. These several complaints about the different professions are often the same for all of them, but there are also some special defects attributed to each. We shall look separately at alleged defects in four professions.

Before doing so, we may ask why it is that complaints about control and responsibility defects in the professions are so numerous and strong just now. There are at least two structural sources of the present complaints. The first lies in the utilitarian interests of the public. The professions are more powerful now than they have ever been and, consequently, they have more power to hurt the public when their control and responsibility mechanisms are not working satisfactorily. The second source is in the changing values of the public, or rather, in the increasingly strong emphasis the public is putting on the old value of equality. The public has less tolerance for professional domination or even paternalistic benevolence. It wishes to participate more fully and more equally in making the decisions that affect its vital interests, both at the level of individual encounter with professionals and at the level of national political decision. In medical therapy and research, for example, patients and subjects wish to participate only on the basis of voluntary and informed consent; with regard to the terms of payment for medical care, they will no longer allow the American Medical Association to dictate what those terms should be. The defects of the powerful professions may be no greater today than in the past, but a new combination of utilitarian and value concerns in the public demands that these defects be reduced or eliminated.

## The Medical Profession

In the public mind, physicians probably constitute the prototype and premier profession. Their knowledge, their prestige, and their power are all considered to be larger than those of other professions. Nonetheless, there are widespread complaints against the medical profession on the grounds both of failures in the realm of service to the public and of defects with regard to effective self-regulation. So far as unsatisfactory service is concerned, the public dislikes the way physicians often seem to be concerned more for science than for caring, to have turned their means into ends, to have become authoritarian and unresponsive, and, finally, to care too much about their money income.

With regard to these several defects, of course, individual physicians differ considerably. But the social control mechanisms by which the best and the average might make the worst come up to higher standards are not very effective. Informal peer controls, as Freidson's research and analysis have shown, do not work well.[12] Physicians who, in group practices, in hospitals, or otherwise, see the faults of their colleagues usually take no action to correct them. They may avoid or even ostracize such colleagues, but these colleagues are then free to commit their harms elsewhere. Those few physicians, nurses, or other paramedicals who go against the system and report incompetent physicians may find themselves silenced or punished. Recently there has been some change of attitude within the profession toward this ineffective informal system of peer control. In New York State, leaders of medical societies are now supporting proposed legislation that would require doctors who learn of acts of professional misconduct to report them to the State Board for Professional Conduct.[13] Under this legislation, failure of a doctor to report "untoward incidents" would itself be professional misconduct that could jeopardize the silent doctor's license. To protect a reporting doctor from legal reprisals, the legislation grants him immunity from civil suits. The sponsors of this legislation are said to feel that it "will lift the veil of silence in the medical profession which protects doctors who are incompetent, ill, or venal and who endanger the health and lives of their patients." Not all physicians are happy about this proposed legislation. In a letter to the editor, one physician calls it "draconian" and a "police-state solution."[14] Since formal rules do not guarantee informal practices, there is no assurance that the legislation will achieve its goal of more effective self-regulation among physicians. Still, it is a sign of a defect that needs a remedy.

Formal control mechanisms established by organizations of physicians have not been much more effective than informal peer control. In each of the local county medical societies into which the American Medical Association is divided for organizational purposes, there exist ethical conduct committees. But, until quite recently, these committees have been minimally effective or even entirely inoperative. And at a still higher level, at the level of what are really the ultimate formal control mechanisms, the state medical licensing boards, which are dominated by the professionals themselves, social control has been weak.[15] It is only with the greatest difficulty that his license can be taken away, even from a physician who is an alcoholic or a narcotic addict.

While there are several sources of the current "malpractice crisis," or the rise in the number of malpractice suits, one of them is undoubtedly just these defects in the informal and formal control mechanisms among physicians. When an aggrieved patient can get no comfort or aid from other physicians, from the ethics committee of the local medical society, or anywhere else in the medical realm, he or she is the more likely to go to the legal system as a last resort. This source of malpractice suits was recognized by the New York State Special Advisory Panel on Medical Malpractice (1976). To keep the risk of medical injury to a minimum, it said, "will require strengthening quality controls over medical practice through a variety of measures, including disciplinary measures for substandard practice, vigorous controls over hospital staff privileges, and limitations of physician practice to areas of proven competence as demonstrated by continuing education and relicensure, if needed."

Medical therapy is not the only part of professional practice that is now being criticized as defective. Medical research using human subjects has also been so criticized, not only in the infamous case of the Nazi doctors, but in the standard procedures of American medical research.[16] Good research evidence shows that at least a significant minority of research physicians has been doing research where risk to subjects exceeded benefits, where satisfactory informed consent was not being obtained, and where the poor and uneducated were more likely to suffer undue risk and be ill-informed than the well-to-do and well-educated. One socially structured source of these defects is the heavier weighting which research physicians give to science as against humane therapy. Another is the lack of education in the ethics of research in the medical schools, in residency training, and in actual clinical practice. There has been some small improvement in this regard, recently, in the form of numerous courses and lectures at the medical schools, but there is still no good evidence that the "scientific stars" who dominate many of these medical schools are treating the ethics of research with the same seriousness with which they take their scientific research itself.[17] The data show that the leading medical centers have not been leaders in concern with the ethics of research as they have been in scientific excellence.[18] Indeed, many medical leaders have expressed indifference or resistance to proposals for reform in the ethics of medical research. They have spoken of "the heavy hand of bureaucracy," of "red tape," even, in some extremes, of "the death of science." New initiatives in this field have come from a few members of the medical establishment,[19] but the chief agent of effective reform has been the government. Perhaps the most powerful agency for reform has been the regulation established by the National Institutes of Health in 1966 that all research it funded thereafter must be approved by a local peer-review committee with regard to satisfactory standards of risk-benefits ratios and of informed consent. More recently, in 1975, under the impetus of the Tuskegee scandal, in which black men were being used as subjects in syphilis research without their informed consent, the Congress established the National Commission for the Protection of the Human Subjects of Biomedical and Behavioral Research. Building on the established N.I.H. regulations, as amended since 1966, the commission has the mandate of recommending new protective procedures for human subjects not only in general but in such specific problem areas as fetal research, psychosurgery, and the use of prisoners and children as subjects. Although both the N.I.H. regulations and the National Commission have been very much influenced by medical professionals, some physicians and researchers still view them as the intrusions of outsiders, an "encroachment of government on medical practice."[20] Such voices are more concerned for the autonomy of the profession than they are for defects in service to the public welfare. Faced with social change, they sound something like businessmen who were faced with similar social changes and "encroachments" of government in the 1930s.

## The Legal Profession

Complaints about defects in public service performance and in effective self-regulation reach a crescendo in the case of the legal profession. Writing a scholarly history of the American legal profession from 1890 to the present, Jerold S. Auerbach entitles his book *Unequal Justice*.[21] In the United States," he says, "justice has been distributed according to race, ethnicity, and wealth, rather than need. This is not equal justice."[22] Nor has this injustice been the fault of the less well trained among the lawyers, the "less professional" members of the bar. "The professional elite," continues Auerbach, "bears a special responsibility for this maldistribution. Its members, absorbed with selective client-caretaking for a restricted clientele, have preserved social and economic inequality. Their efforts, in conjunction with the limitations of an adversary process largely dependent upon the ability to pay, have crippled the capacity of the legal profession to provide equal justice under law or to fulfill those paramount public responsibilities that alone can justify professional independence and self-regulating autonomy."[23] Because his evidence shows in detail that the legal profession does not serve the community as a whole but the interest only of one part—the affluent, the powerful, and the ethnically and racially privileged—Auerbach concludes his book with an eloquent call for more public control and regulation. Auerbach's findings and conclusion are supported by a study of "the unseen power of Washington lawyers," a study that calls these lawyers who sit so close to the national seats of power "the other government."[24] "The bar," *Time* magazine quotes Professor Vern Countryman of the Harvard Law School as saying on the eve of the one hundredth annual meeting of the American Bar Association, "is still dominated by shortsightedness and self-interest."[25]

As with medicine, public distrust of and hostility to the legal profession has expressed itself in a sharply rising rate of malpractice suits.[26] As a result, malpractice insurance rates have risen to such an extent that not only the American Bar Association but "just about every state and local bar association" has set up special committees to consider the "potential crisis in the availability and affordability" of lawyers' malpractice insurance. The rise in the number of legal malpractice suits has occurred not only because of declining public respect for lawyers but also because of changes in the law that make such suits easier and in the willingness of at least a few lawyers to violate the "clubby" collegiality of the law, which tends to define such lawyers as traitors to their profession. In a story on "lawyers who sue lawyers," it is reported that no one yet specializes in such legal malpractice suits, as some lawyers specialize in medical malpractice suits, but now at last there are a few lawyers who will take on a few such cases.[27] In smaller communities, where clubbiness among lawyers is easier to enforce, it is very difficult to find a lawyer willing to take on a legal malpractice suit; in large cities, it is a bit easier. Unfortunately most suits for legal malpractice are settled out of court and with the offending lawyer stipulating, first, that the plaintiff retract his charge of legal malpractice from the record and, second, that the record be sealed to the press and the public. These stipulations prevent effective control of the profession either through public denigration or informal or formal professional sanctions.

Not that such professional sanctions operate very effectively. As with the medical profession, one of the reasons laypersons resort to the courts to complain about lawyers is that there is usually nowhere else they can go with their grievances. Although the Lawyer's Code of Professional Responsibility, which is the American Bar Association's official code of ethics, requires lawyers to report cases of professional misconduct that comes to their attention, few lawyers do so. And the grievance committees of the local bar associations to which such complaints are supposed to be brought have not been well staffed, active, or effective in investigating them when they are brought directly by laypeople. The offending lawyers have been permitted to use all the forms of legal evasion granted under generous interpretations of due process. In any case, it seems to be only the "small fry," the "less professional" among the lawyers who are ever even brought up before the grievance committees. Quite recently, under the influence of increasing protests from the public and from a small group within the profession itself, especially the younger, "public interest" lawyers, the grievance committees have been making small efforts to reform themselves. The bar associations have granted them more funds to hire more staff and have urged them to do their job more effectively. In some cases, however, the local bar associations have not been willing to go against the particularistic clubbiness that prevails, especially in the establishment bar, and have passed their control responsibilities on to other agencies. In Illinois, for example, the Chicago Bar Association and the Illinois State Bar Association supported the setting up, in 1973, of the Attorney Registration and Disciplinary Commission by the state's Supreme Court to be responsible for professional self-regulation.[28] The Chicago and state bar associations successfully opposed the membership of laity on this new commission. Thus, clubbiness is perpetuated, though it is likely that the new commission, with its larger and more professional staff, its visibility and legitimacy, and its removal from the immediate local community, will be more effective than the previous disciplinary agencies.

There remain those who know the legal profession and who doubt that anything less than public disciplinary control with laypersons participating will be really effective in maintaining high professional standards. Monroe Freedman, former dean of the law school at Hofstra University and the author of a textbook on legal ethics, says that effective control is not going to come from the bar associations. "The bar," he says, "has never policed itself adequately . . . . there's too much self-protection there . . . and too much betraying of a client's interests."[29] A public-interest group sponsored by Ralph Nader takes the same position. This group, called Public Citizen, in a 186-page study of the disciplinary procedures of the bar associations in Baltimore, Boston, New York, Philadelphia, and Washington, concludes that only the continuous participation of and pressure from informed citizens will make the legal profession produce better and cheaper service.[30]

*The Accounting Profession*

Accounting is one of the modern professions, not so old or prestigious as medicine or law, but growing in power and consequentiality for the public welfare. Accounting has to do with setting and checking standards of information and accountability in the financial statements of public and private organizations. The 130,000 accountants who are members of the American Institute of Certified Public Accountants and the large firms in which many of them work, are responsible, for example, for the reliability of the financial information issued by American corporations, information on which individual investors, pension funds, and various institutions depend to make invest-ment decisions. If this information is unreliable, investors may lose hundreds of million of dollars, as has happened in some notably scandalous cases de-tailed by Briloff.[31] Accountants are also responsible for an "independent au-dit" or certification of the financial information issued by these corporations as the basis on which the government assesses taxes. Unreliable and exces-sively flexible accounting practices and procedures may cost the government and the public, again, hundreds of millions of dollars. It is obvious that ac-counting is very much and increasingly affected with the public interest. It is also a profession that is increasingly based on the generalized knowledge created by academic economics and the other social sciences. There is, of course, a range of variation within the profession, from mere "bookkeepers," who perform only the simplest of accounting tasks, to academic professors who are responsible, in some cases, for advancing the theory and standards of accounting.

How well does the accounting profession measure up on the dimensions of public-service performance and effective self-regulation? Not very well, ac-cording to a recent 1,760-page report of the Committee on Government Oper-ations of the U.S. Senate[32] and to such critics as Professor Briloff.[33] Accord-ing to the Senate report, the accounting profession is dominated by a few very large firms, the so-called "Big Eight." These firms dominate the profession through their control of the committees, offices, and staff of the key profes-sional organizations, such as the American Institute of Certified Public Ac-countants, the Financial Accounting Foundation, and the Financial Account-ing Standards Board, which set standards for the profession. Though they are supposed to be "independent" professional auditors of the financial state-ments of America's corporations (the "Big Eight" audit the accounts of 85 percent of the corporations listed on the New York Stock Exchange), these firms often seem to care more about the interests of the corporations than the public welfare. They identify with the corporations rather than the public and join them in lobbying efforts against congressional efforts to reform account-ing and corporate reporting standards. Through their domination of the ac-counting standards boards, the "Big Eight" practice what they call "creative accounting," that is, they adopt standards that have been labeled by outsiders as being so flexible that corporations sometimes seem to be able to call profits losses, and losses profits, depending on which label is most profitable and least taxable. In a considerable number of cases that have later been publicly exposed, accounting firms have aided corporations in deceiving the public and the government.

Among the proposals for reform of this unsatisfactory situation recom-mended in the Senate report were stronger oversight of accounting practices by Congress, the direct establishment of financial accounting standards by the government itselt through a special commission for that purpose, and "participation by all segments of the public" in these tasks. When committee hearings on the report were held, however, the senators finally abstained from recommending any legislation toward these ends at the time. Responding to proposals from the accounting profession's leaders for self-reform, the senators decided to give the profession another chance to make itself more "professional."[34]

## The Academic Profession

Though purists among them sometimes resist being so labeled, academic scientists and scholars are "professionals" in high degree by the several criteria specified in our definition. They are the creators and teachers of the most highly generalized knowledge. And they have large effects on the public welfare in a number of ways.[35] They mold the values and ideologies of their undergraduate students; they train the other professionals; their generalized knowledge is essential for maintaining and advancing performance in the practicing professions, thus also indirectly affecting the general public; and they often even have important direct effects on public values and welfare, as when they discover atomic energy, do experiments on human subjects, do recombinant DNA research, or, as writers of school textbooks or as intellectuals, discourse on the nature of man and on man's evolution from animal species. Thus, academic scientists and scholars are a profession by virtue of both their generalized knowledge and their consequentiality for the public welfare. There is, of course, a range of "professionality" among them, extending from routine teachers in marginal colleges to distinguished "stars" in the sciences and the humanities in the great universities.

Even here, however, we find less than perfectly satisfactory service in the public welfare and occasional lapses from effective self-regulation. Means become ends even for academic professionals.[36] For example, in their zealous pursuit of discoveries, academics become heedless of the consequences of those discoveries for the public welfare and of the costs of some of their procedures to the individuals who are the objects and subjects of their research. A great deal of the recent outcry against the abuses of human experimentation by biomedical scientists is against practices located to a considerable extent in premier academic institutions.[37] And when the National Institutes of Health, in its guidelines designed to curb abuses of this kind, required lay participation on the local institutional ethical review boards, this requirement met with hostility in many academic quarters. Academic scientists have done somewhat better about their public responsibility and self-regulation in the case of the possible dangers resulting from recombinant DNA research by taking the initiative, at Asilomar in 1975, of setting up protective standards for all such research. But the same scientists have been at least ambivalent and sometimes hostile when local or national governmental regulations for the dangers of this research have been proposed. Scientists and scholars sometimes assert claims to autonomy that are viewed as unwarranted and "arrogant" by some sections of the public. In the fluoridation controversy, for example, scientists were denounced by political rightists for imposing their views on the public.[38] In the case of the controversy over building nuclear energy plants, the charge is the same, though this time it is made by political leftists. In her study of the science-textbook controversies where there was public criticism of the views on evolution stated in biology and social-science textbooks, Nelkin points out that for these critics science has become a "symbol of an authoritarian ideology." These critics "express extraordinary resentment of 'scientific dogmatism,' of the 'arrogance' and 'absence of humility' among scientists." Criticism of "the dominance of scientific values and the role of expertise, as well as demands for increased local participation," she continues, "can be found among people of a wide spectrum of political ideologies. . . . All these groups have expressed disillusion with technology and expertise, and their slogans are 'accountability,' 'lay participation,' and "demystification of expertise.'"[39]

The public is not radically anti-intellectual or antiscientist, as some academics fear, but it is ambivalent about the mixed consequences of academic professionalism and it is, increasingly, demanding more lay and democratic control over some of these consequences. Academic professionals who resent or resist an increased measure of public control may well take note of Duncan Macrae's remark: "Democracy, however, requires that the electorate have the ultimate power. Those who value democracy, or fear its erosion, sometimes see scientists as an elite serving special interests, or see science as simply unplanned and uncontrolled."[40]

## New Mechanisms for Improved Control and Responsibility in the Powerful Professions

Social change is often hard to come by, and slow. Nonetheless, along with the widespread investigation and criticism of present control and responsibility processes in the powerful professions, there have been many moves toward reform and greater professionalism from both inside and outside the professions, sometimes separately, but also frequently conjointly. Individual professionals, voluntary groups, the foundations, and the state and federal governments have all contributed, often together, in these reform efforts.

Fortunately for professional standards and for the public welfare, there are always a few members of each profession who are dissatisfied with its present performance and who become advocates of reform from within. Among biomedical researchers, Professor Henry Beecher, very much a member of the establishment in that group, became a powerful agent of reform with his exposure of widespread abuses of the rights of human subjects.[41] As with some other criticisms from within the professions, Beecher's was somewhat weakened by his tendency to see these abuses as the faults of "bad guys" in the profession, not the clear result of defects in the existing system of self-regulation.[42] But internal critics often see these system defects. For example, a group of young lawyers in the Chicago Bar Association, in order to reform the CBA's inadequate system of control over unprofessional lawyers, formed the Chicago Council of Lawyers to bring about needed changes in the system. Powell points out that "the Council was not interested in patching up the existing system; it was clearly beyond repair and was inherently inadequate, controlled as it was by lawyers without reference to members of the public."[43] Eventually the council was successful in having established a better system, a statewide Attorney Registration and Disciplinary Commission. System defects in the present processes of legal practice and legal education have also been pointed out by Douglas Rosenthal[44] and Jerold Auerbach.[45] Rosenthal's research, done for a Ph.D. at Yale, shows that the typical authoritarian stance that lawyers take toward their clients is no more successful in winning suits than a stance that actively includes the lay client in the prosecution of his case. Rosenthal is a practicing attorney in the U.S. Justice Department. And Auerbach, trained as a lawyer but now a professional historian, indicts the profession for its preference for the powerful in American society and its lack of equal service to the poor and less powerful.

Insider professionals do not limit themselves to criticisms and to proposals for reform. They have been active in setting up agencies for reform of professional obligations. With the financial help of private individuals and the foundations, notably the Ford Foundation,[46] young, liberal members of a number of professions have set up public-interest professional organizations to serve the public interest directly and continuously. Public-interest law firms have been active in such areas, where there are important public interests that have not been adequately represented before, as the environment, health services, housing, employment, communications, education, and women's and minority rights.[47] What the lawyers started, economists and tax experts and accountants have recently continued. The Ford Foundation has supported the establishment of Tax Analysts and Advocates, the National Association of Accountants in the Public Interest, and the Public Interest Economics Foundation.[48] Finally, with support from their own members and from other sources, academic scientists have established a number of public-interest organizations to provide knowledge and guidance on public issues where science matters were important.[49] While most of these are for scientists only, a few include laity. For example, an organization called Public Responsibility in Medicine and Research has been established in Massachusetts "as a vehicle by which researchers, clinicians, research and health care administrators, attorneys, and members of the lay public can be brought together" in response "to the increasingly complex and sensitive problems currently facing research and related clinical practice."[50]

Nonetheless, all this has not been enough. Establishments, among professionals as among other social groups, are always conservative when what they define as their central values are being subverted, as they see it. The professions have been resistant to both insider and outsider attacks on their absolute autonomy. As a result, outside reforms, especially from a variety of government actions and agencies, have been necessary to supplement and strengthen insider initiatives for reform.

In biomedical research, for example, Beecher's criticisms of abuses of human subjects would not have been effective alone. Medical schools did not set up local peer-review committees for the control of such abuses until the National Institutes of Health, on whom they depended vitally for financial support of their research, required them in 1966 to set up such committees on pain, otherwise, of not being funded.[51] Administrative action of this kind by the National Institutes was supplemented in 1973 by legislative action by the Congress in establishing the National Commission for the Protection of the Human Subjects of Biomedical and Behavioral Research. The commission was set up, after the Tuskegee scandal showed that abuses were still occurring, to improve the treatment not only of subjects in general but of certain special classes of them: children, prisoners, and those receiving psychosurgical interventions. In addition to these actions by the government, there have been investigatory commissions that have looked into defects in the medical profession, as in the case of malpractice.[52] Also, as we have seen, various state governments have taken action. The New York State Board of Regents, going against established professional codes, has allowed not only doctors and dentists but also a variety of other professionals whom it regulates to advertise, under specified conditions, to the public.[53] Advertising alone, as Ladinsky has pointed out, will be of limited usefulness in achieving more lay control over the medical profession, but it is a step in that direction.[54]

Not only state boards, but the Supreme Court itself has decided, on petition from a couple of lawyers, that professional codes which prohibit advertising by lawyers are illegal.[55] In its 1977 annual meeting, the American Bar Association, with great reluctance, acceded to this judgment and has formulated rules for advertising by its members.[56] So far as the legal profession is concerned, the federal government has gone beyond mere judgments to definite action to correct defects in public service responsibilities. The Legal Services Corporation of the government's Office of Economic Opportunity is now operating on a budget of $200 million in its provision of legal services to the poor who would not otherwise get them.[57]

## Concluding Comment

For the future, we can look to continued efforts for the improvement of professionalism in the powerful professions through a mixture of better self-regulation, continued public concern, and stronger government actions where self-regulation is not sufficient. Professional activities are clearly too important to the public welfare to be left to the control of the specialized professionals themselves. Democratic and egalitarian sentiments in American society will continue to be in tension with the inherent elitism and power of professional knowledge and the associations which build up around it. Those who believe in rational remedies for social problems of this kind will hope that the social sciences will be able, through research and analysis, to make their contribution to constructive accommodations and resolutions of the tensions between egalitarianism and elitism. That tension will not go away. Our best hope is for improving solutions that pay their full respects to both egalitariansim, on the one hand, and specialized and powerful and useful knowledge, on the other.

## Notes

1. *New York Times*, April 29, 1977.
2. *New York Times*, July 22, 1977.
3. U.S. Department of Health, Education, and Welfare, *Report of the Secretary's Commission on Medical Malpractice* (Washington, D.C.: U.S. Government Printing Office, 1973).
4. State of New York, *Report of the Special Advisory Panel on Medical Malpractice*. Albany, N.Y., 1976.
5. U.S. Senate, Committee on Government Operations, Subcommittee on Reports, Accounting and Management, *The Accounting Establishment* (Washington, D.C.: U.S. Government Printing Office, 1976).
6. See Dorothy Nelkin, *Science Textbook Controversies and the Politics of Equal Time* (Cambridge, Mass.: MIT Press, 1977); Dorothy Nelkin, "Scientists and Professional Responsibility: The Experience of American Ecologists," *Social Studies of Science* 7 (1977): 75–95; Joseph Ben-David, "Science as a Profession and Scientific Professionalism," in J. J. Loubser et al., eds., *Explorations in General Theory: Essays in Honor of Talcott Parsons* (New York: Free Press, 1976); Allan Mazur, "Disputes between Experts," *Minerva* 11 (1973): 243–62; and Allan Mazur, "Opposition to Technological Innovations," *Minerva* 13 (1975): 58–81.
7. M. R. Haug and M. B. Sussman, "Professional Autonomy and the Revolt of the Client," *Social Problems* 17 (1969): 153–61; and M. R. Haug, "The Deprofessionsalization of Everyone?" *Sociological Focus* 8 (1975): 197–213; and M.R. Haug, "The Erosion of Professional Autonomy: A Cross-cultural Inquiry in the

Case of the Physician," *MMFQ/Health and Society* (Winter 1976): 83–106.

8. On the theory of the professions, see E. Durkheim, *Professional Ethics and Civic Morals*, trans. C. Brookfield (London: Routledge & Kegan Paul, 1957); T. Parsons, "The Professions and Social Structure," *Social Forces* 17 (1939): 457–67; Bernard Barber, "Is American Business Becoming Professionalized? Analysis of a Social Ideology," in E. A. Tiryakian, ed., *Sociocultural Theory, Values, and Sociocultural Change: Essays in Honor of P. A. Sorokin* (Glencoe, Ill.: Free Press, 1963); Bernard Barber, "Some Problems in the Sociology of the Professions" *Daedalus* (Fall 1963): 669–88; Bernard Barber, "Compassion in Medicine: Toward New Definitions and New Institutions," *New England Journal of Medicine* 295 (1976): 939–43; W. E. Moore, *The Professions: Roles and Rules* (New York: Russell Sage Foundation, 1970); Nathan Reingold, "Definitions and Speculations: The Professionalization of Science in America in the Nineteenth Century," in A. C. Oleson and S. Brown, eds., *Knowledge in the Early American Republic* (Baltimore, Md.: Johns Hopkins University Press, 1976); and Harold Wilensky, "The Professionalization of Everyone?" *American Journal of Sociology* 70 (1964): 137–58.

9. Bernard Barber, *Social Stratification* (New York: Harcourt Brace, 1957), and Bernard Barber, "Inequality and Occupational Prestige: Theory, Research and Social Policy," *Sociological Inquiry*, 48 (2)(1978): 75–88.

10. Amitai Etzioni, *The Semi-Professions and Their Organization* (New York: Free Press, 1969).

11. J. L. Berlant, *Profession and Monopoly* (Berkeley: University of California Press, 1975).

12. Eliot Freidson, *Professional Dominance* (New York: Atherton Press, 1970); Eliot Freidson, *Profession of Medicine* (New York: Dodd, Mead, 1970); and Eliot Freidson, *Doctoring Together* (New York; Elsevier, 1975).

13. *New York Times*, April 29, 1977.

14. *New York Times*, June 11, 1977.

15. H. Lewis and M. Lewis, *The Medical Offenders* (New York: Simon & Schuster, 1970), part I, "The Crisis in Medical Discipline."

16. Jay Katz, ed., *Experimentation with Human Beings* (New York: Russell Sage Foundation, 1972); Bernard Barber et al., *Research on Human Subjects* (New York: Russell Sage Foundation, 1973); B. H. Gray, *Human Subjects in Medical Experimentation* (New York: Wiley-Interscience, 1975); and S. J. Reiser et al., eds., *Ethics in Medicine* (Cambridge, Mass.: MIT Press, 1977).

17. J. J. Lally, "The Making of the Compassionate Physician-Investigator," *Annals* 437 (1978): 86–98. This issue of *Annals* is a special one on *Medical Ethics and Social Change*, Bernard Barber, special ed.

18. Barber et al., *Research on Human Subjects*.

19. H. K. Beecher, *Research and the Individual* (Boston: Little, Brown, 1970); S. R. Graubard, ed., "Ethical Aspects of Experimentation with Human Subjects," *Daedalus* (Spring, 1969).

20. I. H. Page, "What Price Protection?" *Journal of the American Medical Association* 235 (1976): 286; I. H. Page, "Ecumenicalism in Medicine and Science—Its Urgent Needs," *New England Journal of Medicine* 297 (1977): 215–16.

21. Jerold S. Auerbach, *Unequal Justice: Lawyers and Social Change in Modern America* (New York: Oxford University Press, 1976).

22. Ibid., p. 12.

23. Ibid.

24. M. J. Green, *The Other Government: Unseen Power of Washington Lawyers* (New York: Grossman (Viking), 1975).
25. *Time*, August 8, 1977.
26. See, e.g., *New York Times*, February 28, 1977; and *Wall Street Journal*, August 17, 1977.
27. *New York Times*, June 26, 1977.
28. M. J. Powell, "Professional Self-Regulation: The Transfer of Control from a Professional Association to an Independent Commission," American Bar Association, mimeo.
29. *New York Times*, June 26, 1977.
30. *New York Times*, July 18, 1977.
31. A. J. Briloff, *More Debits Than Credits: The Burnt Investor's Guide to Financial Statements* (New York: Harper & Row, 1977).
32. U.S. Senate report, *The Accounting Establishment*.
33. Briloff, *More Debits Than Credits*.
34. *New York Times*, April 1, 12, 20, June 14, 20, 1977.
35. Rae Goodell, *The Visible Scientists* (Boston: Little, Brown, 1977).
36. Robert K. Merton, *The Sociology of Science* (Chicago: University of Chicago Press, 1973), p. 262. This essay was originally published in 1938.
37. Jay Katz, *Experimentation*; Barber et al., *Research on Human Subjects*; and Gray, *Human Subjects*.
38. Allan Mazur, "Disputes between Experts," and "Opposition to Technological Innovation."
39. Nelkin, *Science Textbook Controversies*, pp. 131, 138.
40. Duncan Macrae, Jr., "Science and the Formation of Policy in a Democracy," *Minerva* 11 (1973): 228–42.
41. Beecher, *Research and the Individual*; Beecher's English counterpart is M. H. Pappworth, *Human Guinea Pigs* (London: Routledge & Kegan Paul, 1967).
42. John J. Lally and Bernard Barber, "The Compassionate Physician: Frequency and Social Determinants of Physician-Investigator Concern for Human Subjects," *Social Forces* 53 (1974): 289–96; Barber, "Compassion in Medicine."
43. M. J. Powell, "Professional Self-Regulation."
44. Douglas E. Rosenthal, *Lawyer and Client: Who's in Charge?* (New York: Russell Sage Foundation, 1974).
45. Auerbach, *Unequal Justice*.
46. Ford Foundation, *The Public Interest Law Firm: New Voices for New Constituencies* (New York, 1976).
47. Ford Foundation, *Newsletter*, June 1, 1977.
48. Ibid.
49. Nelkin, *Science Textbook Controversies*; Nelkin, "Scientists and Professional Responsibility"; M. Perl, "Public Interest Science," *Physics Today*, June 1974; J. Primack and F. von Hippel, *Advice and Dissent* (New York: Basic Books, 1974); and J. M. Berry, *Lobbying for the People* (Princeton: Princeton University Press, 1977).
50. *PRIM&R*, a newsletter, January 1977.
51. Barber et al., *Research on Human Subjects*.
52. U.S. Dept. of Health, Education, and Welfare, *Report . . . On Medical Malpractice*.
53. *New York Times*, July 29, 1977.
54. Jack Ladinsky, "The Traffic in Legal Services: Lawyer-Seeking Behavior and the

Channeling of Clients," in L. Brickman and R. Lempert, eds., *The Role of Research in the Delivery of Legal Services* (Washington, D.C.: Resource Center for Consumers of Legal Services, 1976).

55. *New York Times*, June 28, 1977.
56. *Wall Street Journal*, August 17, 1977.
57. *New York Times*, July 31, 1977.

Part IV

# Relations between the Sociology of Science and Related Scholarly Fields

# Introduction

Science as a whole is not one integrated and harmonious whole. The relationships among disciplines and among specialties within the disciplines are complex and diverse. In addition to synergistic relationships, there are those that are filled with tension and conflict. I have remarked in the introduction to this entire volume on the tension and conflict that existed between the history of science and the sociology of science when the historians believed in the internalist-externalist dichotomy. That conflict is gone now and the two specialties are fruitfully interacting. Indeed, there is now a synergistic relationship among the history, philosophy, and sociology of science. Here in part IV, three papers discuss some of these complex relationships between the sociology of science and other specialties.

The first paper (chap. 15), "On the Relations between Philosophy of Science and Sociology of Science," written at the request of the Program Committee of an annual meeting of the Philosophy of Science Association that was devoted to assessing the current state of that field, traces the changing relationships between the sociology and philosophy of science over a forty-year period. As sociology moved away from its fear of philosophy, and philosophy moved away from its predominant emphasis on language and logic, they came together in their present mutually beneficial situation. Drawing on social-systems theory for help, I make suggestions, as requested by the Program Committee, for what the relationships between these two fields should be.

The second paper (chap. 16), "Scientists and the Social Study of Science: A Research Problem," was written as my Presidential Address in 1979 to the Society for Social Studies of Science (4–S). Starting with my expression of dismay at the widespread ignorance among natural scientists of work in the sociology of science, their seeming denial that such is even possible, and their public assertion of all kinds of sociological, political, and psychological generalizations about scientists and their work without any awareness of the limitations of their impressions, their prejudices, and their commonsense knowledge, I suggest some hypotheses and research possibilities to discover whether and why these patterns exist. Obviously, more cooperation among natural and social scientists is called for.

Finally, in the last paper (chap.17), "Tension and Accommodations between Science and Humanism," I offer my analysis of both the sources of tension and the patterns of accommodation between science and humanism, a matter that has, ever since C. P. Snow wrote on this topic, been called "the two-culture problem."[1] It is unlikely that this tension will soon disappear, but harmful conflict can certainly be diminished.

## Note

1. C. P. Snow, *The Two Cultures and the Scientific Revolution* (Cambridge, England: Cambridge University Press, 1959). This brief volume, containing Snow's Rede Lectures, had a very large effect on the discussion of its topic.

# 15

## On the Relations between Philosophy of Science and Sociology of Science

My topic concerns what the relations between the philosophy of science and the sociology of science *have* been and what they *ought to be*. In answer to the first question, what these relations have been, I shall rely on a paucity of scholarship on this subject and, mostly, on my own impressions and experience during the past forty years or so, experience that began when I was an undergraduate studying under Talcott Parsons, Robert Merton, P. A. Sorokin, George Sarton, and L. J. Henderson, all of whom, in their various ways, were important figures in the sociology, history, and philosophy of science. In answer to the second question, what these relations ought to be, I shall refer to a systematic theoretical model and partly to actual work presently going on in these two fields.

What, then, have the relations between the philosophy and sociology of science been these many years? You must remember that during this period sociology as a whole has been struggling to establish itself as a valid and accepted science. Sociology as a social movement has lusted after science, and many of its present virtues and vices are due to this passion. Moreover, this new passion required its devotees to reject what were defined as their old loves, the evils that had seduced them into false paths. One of these evil old loves, of course, was something called philosophy, not necessarily just what a technical philosopher would mean by philosophy, but a whole variety of vague, speculative, nonempirical, overgeneralized assumptions and methodologies that attracted the founding fathers of American sociology, especially those who had been influenced by Germanic education and models. Recent sociology has been out to establish its independence of philosophy, its independence as a separate and scientific discipline.

In none of the subfields of sociology as a whole has this effort to break away from philosophy been stronger than in the sociology of science. This was because the sociology of science started off by being defined as part of the larger field of *Wissenssoziologie*, or sociology of knowledge; and as the German name for the field indicates, it was very much a field in which Germanic concerns and predispositions for the philosophical, the generalized, the speculative, and the nonempirical prevailed. To this day, in my graduate course in the sociology of knowledge, I have to ward off students who come with expectations and desires for this Germanic model (Barber [3]). I still have the students read Mannheim's *Ideology and Utopia* [12], one of the founding classics of the sociology of knowledge, but I also have them read Merton's admiring exposition and critique [13]. Merton's admiration, of course, is for the sociological elements in Mannheim's analysis, his attempts to show how certain social or cultural variables, e.g., ideas about history, were related to other social or cultural variables, e.g., the particular generation to which the holders of these ideas belonged. His critique was for the philosophical confusion in Mannheim, the inability or unwillingness to establish some firm rational rules for justifying the validity of scientific knowledge.

Sociologists of science have focused, then, on the sociological problems, the hard tasks of relating one social or cultural variable or set of such variables to specified other social or cultural variables or sets of them (Barber [1], [2]; Merton [15]). We have been willing, indeed I have insisted to my students, that philosophical problems were special kinds of technical problems that should be left to the relevant experts, the philosophers of science, like Kaplan or Nagel or Popper or Lakatos. Although I have always said that there were important interactions between sociology and philosophy which ought to be studied, until quite recently I also said that I did not think there existed any philosopher of science who knew enough about sociology or any sociologist of science who knew enough about philosophy to make such study worthwhile. What I tell them now I shall reveal in a little while when I say more fully what the relations between the philosophy and sociology of science ought to be. I should note that the sociology of science, especially in the case of pioneers like Merton and me, who came under the direct influence of historians of science, was more receptive to the history of science. Merton's *Science, Technology and Society in Seventeenth Century England* [14] is, of course, a classic of sociological history. In my *Sociology of Science* reader [2], I included not only excerpts from Merton's work but pieces by historians such as Richard Shryock on American science and Charles Gillispie on French science. Indeed, in the early 1960s, when Tom Kuhn, an old friend of ours, was trying to get the University of Chicago Press to publish his *The Structure of Scientific Revolutions* [8] in hardcover as well as in the softcover *International Encyclopedia of Unified Science*, he asked us to write supporting letters. One of the happy recent developments in the history of science, from our point of view, has been the much greater appreciation and use of sociological concepts and methods. Just as examples, I might mention Roger Hahn [7], Arnold Thackray (Shapin and Thackray [16]) and Roy MacLeod [11].

Gradually, very gradually, the sociology of science has achieved something of the scientific status it has longed for. Cole and Zuckerman [6] have recently shown, using citation data, the great increase in cognitive consensus in the field from the early period 1950–54 to the period 1970–73. And besides a decent amount of cognitive consensus, the sociology of science has achieved all the essential characteristics of an institutionalized scientific field: regularized university courses of instruction, special journals, special funding agencies, special professional associations, and specialized scholarly conferences. Perhaps the time has come when the sociology of science has sufficient security and acceptance as a scientific specialty that it can afford to look more knowledgeably and less fearfully at the philosophy of science. Though some recent British sociology of science writing (Barnes [4], Blume [5]) seems to imply otherwise, sociological relativism does not require the kind of philosophical relativism that would subvert the sociology of science as a viable scientific enterprise.

Now let me turn to a systematic answer to the second question, what the relations between philosophy and sociology of science should be. My answer is derived from a systematic theoretical model I have constructed for the analysis of "action" in social systems. Both the philosophy and sociology of science are specialties for the analysis of "action" in social systems. As Figure 15.1 may help to make clear, I assume that "action" or "interaction" is the basic stuff studied by the social science and humanistic disciplines, just as "matter" is for physics and "life" is for the biological sciences. Figure 15.1 shows that I assume further that the endless process of action or interaction is usefully conceived, not necessarily ontologically but only for purposes of scientific analysis, to be divided into separate, boundary-maintaining, dynamic, and changeful systems. One type of system that is useful as a comprehensive framework for social science and humanistic analysis is the one we label "society," but the system assumption is useful also for the elements that are constituents at various levels and sublevels of the society model.

| SOCIAL STRUCTURE | CULTURE | PERSONALITY |
|---|---|---|
| BASIC UNIT—STRUCTURED AND PROBABLY INSTITUTIONALIZED "ROLE" | BASIC UNIT—STRUCTURED AND PROBABLY INSTITUIONALIZED "IDEA" | BASIC UNIT—"?" |

*Figure 15.1.* Action (Interaction), Social Systems, Society.

Figure 15.1 also illustrates the assumption of this model that all action in social systems has three analytic aspects: social structure, culture, and personality. That is to say, there is no concrete unit of action in any social system, societal or smaller, in which all three of these aspects of its structure and functioning cannot be analytically discerned. As Figure 15.1 shows, the basic unit of analysis for social structure is "role" and for culture is "idea" (or "symbol" will do). Because of paradigm differences among schools of psychologists, who are the disciplinary experts with regard to personality, the basic unit of analysis for that aspect of action is left unspecified in this model.

All three aspects of action are conceived of as structured (or patterned, a synonym). That is, there are assumed to be discernible uniformities of regularity and recurrence in the process. Structure and process are but two ways of conceiving the basic stuff of action. Figure 15.1 also indicates that role and idea structures are to be analyzed as more or less "institutionalized," that is, as supported by more or less moral consensus among those who enact and maintain these structures. Institutionalization is very much a matter of degree, with some structured systems having little moral consensus attached to them, others having much more. According to this model, idea structures are subject to the same assumptions and analysis as role structures. Language, ideologies, science, philosophy, and all the other types of idea-system are analyzable not only in terms of their cognitive or logical structures but as foci of sentiments of moral consensus and dissensus among those who use and develop them. Further, Figure 15.1 is intended to show that all three aspects of action—social structure, culture and personality—are independent of one another to some specifiable extent, as well as being interdependent in some specifiable ways and degrees. That is to say, in some ways their essential functions, types of structure, dynamics, and change are different from one another. These differences are matters for theoretical analysis and empirical research combined. Finally, Figure 15.1 assumes that social structure, culture, and personality are *in principle* all equally important aspects of action. In concrete cases of analysis, of course, any one may be more important than the others, but that is something to be established empirically and by controlled or natural experimentation. The model presented in Figure 15.1 rejects all the monofactorial theories—whether social-structural, cultural, or personality—which are rife in the social and humanistic disciplines today.

As Figure 15.2 shows, each of the three systemic aspects of action can be divided into a number of functional subsystems. All the assumptions that apply to the inclusive categories apply to these subsystems. They are all structured, institutionalized to some degree, somewhat independent of one another but also interdependent and, finally, in principle all equally important. Both science and philosophy are cultural subsystems, necessarily in relationship

| | |
|---|---|
| **SOCIAL STRUCTURE** | SOCIALIZATION (EDUCATION) STRUCTURE |
| | ECONOMIC STRUCTURE |
| | POLITICAL STRUCTURE |
| | STRATIFICATION STRUCTURE |
| | KINSHIP STRUCTURE |
| | COMMUNICATIONS STRUCTURE |
| | ORGANIZATION (GROUP) STRUCTURE |
| **CULTURE** | IDEOLOGY (STRUCTURED OR PATTERNED) |
| | SCIENCE (STRUCTURED OR PATTERNED) |
| | RELIGION (STRUCTURED OR PATTERNED) |
| | VALUES (STRUCTURED OR PATTERNED) |
| | PHILOSOPHY (STRUCTURED OR PATTERNED) |
| | LANGUAGE (STRUCTURED OR PATTERNED) |
| | ART (STRUCTURED OR PATTERNED) |
| **PERSONALITY** | |

*Figure 15.2.* Society.

not only with one another but each also with a variety of other cultural, social-structural, and personality systems. The task of the sociology of science is to analyze and study not only the independent functions and structures of science in its various forms, natural, biological, and social, but also its interdependence with social-structural elements such as class systems, political systems, and organizational forms and with cultural elements such as philosophy, religion, and art. The task of philosophy as a discipline is to study the solutions provided by professional philosophers and others to a variety of problems of meaning in the ontological, epistemological, and moral spheres. Perhaps also it is to study the historical and social, cultural, and personality interdependencies of philosophy. But this second task, especially insofar as these interdependencies are concerned, has been much less esteemed and developed in the discipline of philosophy than the former. Once these interdependencies are acknowledged, as they are in the model I have presented, they might become a more honored and cultivated occupation for professional philosophers than they now are. Philosophers of science have done a good deal toward learning the substance of the physical sciences in order to study the interdependency of science and philosophy, though they have tended to look more at how philosophy, influences science than the other way around. They have done much less with the biological and social sciences; perhaps they will do more in the future. If they do, they will have to learn something of the technical substance of these fields, in order to respect their degree of independence, as well as to bring their own technical expertise to bear.

As for sociologists of science, because of the interdependency of science and philosophy, more of them ought to learn the technical substance of philosophy in order to study this interdependency. I do not see much movement in this direction among American sociologists of science. The British and German sociologists of science, more of whom have studied philosophy in the university or have come into sociology from philosophy, have done more in this respect, though for the present they seem too enamored of a philosophical relativism.

One hopeful sign is that in recent writings that have been defined as important for the philosophy of science a variety of sociological notions are employed. In Kuhn, for example, notions of scientific communities, of competition between schools, of social and cultural resistance to innovation, and of social processes of scientific training and mentorship are all present, though often implicitly. In Lakatos [9], again, despite his occasional overt scorn for the social sciences, there are a number of sociological notions. Just as we need sociologists of science who will learn some philosophy to achieve better work in the interdependent relationships between the two, so we need philosophers of science who will not just make up their own sociology of science as they go along but will attend to the technical conceptual and methodological apparatus of established sociology of science for the same purpose. For the present we may conclude with Lakatos's paraphrase [10] of Kant's famous dictum, expanding Lakatos only to include the sociology of science: "Philosophy of science without history of science is empty; history of science without philosophy of science is blind." All three disciplines, because of the interdependency between science and philosophy, must pay the closest and most expert attention to one another.

# References

1. Barber, Bernard. *Science and the Social Order*. Glencoe, Ill.: Free Press, 1952.
2. Barber, Bernard, and Walter Hirsch, eds. *The Sociology of Science*. Glencoe Ill.: Free Press, 1962.
3. Barber, Bernard. "Toward a New View of the Sociology of Knowledge." In *The Idea of Social Structure: Papers in Honor of Robert K. Merton*. Ed. Lewis A. Coser. New York: Harcourt, Brace, Jovanovich, 1975. pp. 103–16.
4. Barnes B. *Scientific Knowledge and Sociological Theory*. London and Boston: Routledge & Kegal Paul, 1974.
5. Blume, S. S. *Toward a Political Sociology of Science*. New York and London: Free Press, Macmillan, 1974.
6. Cole, Jonathan R., and Harriet Zuckerman. The Emergence of a Scientific Specialty: The Self-Exemplifying Case of the Sociology of Science." In *The Idea of Social Structure: Papers in Honor of Robert K. Merton*. Ed. Lewis A. Coser. New York: Harcourt, Brace, Jovanovich, 1975. pp. 139–74.
7. Hahn, Roger, *The Anatomy of a Scientific Institution: The Paris Academy of Sciences. 1666–1803*. Berkeley: University of California Press, 1971.
8. Kuhn, Thomas. *The Structure of Scientific Revolution*. Chicago, Ill.: University of Chicago Press, 1962.
9. Lakatos, Imre. "Falsification and the Methodology of Scientific Research Programmes." In *Criticism and the Growth of Knowledge*. Ed. I. Lakatos and A. Musgrave. Cambridge, England: Cambridge University Press, 1970. pp. 91–195.
10. Lakatos, Imre. "History of Science and Its Rational Reconstruction." In *PSA 1970 (Boston Studies in the Philosophy of Science*. vol. 8). Ed. R. S. Cohen and R. C. Buck. Dordrecht: D. Reidel, 1971. pp. 91–136.
11. MacLeod, Roy. "Changing Perspectives in the Social History of Science." In *Science, Technology, and Society*. Ed. I. Spiegel-Rosing and D. Price. London and Beverly Hills: Sage Publications, 1977. pp. 149–95.
12. Mannheim, Karl. *Ideology and Utopia*. Trans. E. Shils and L. Wirth. New York: Harcourt Brace, 1946. (Originally published as *Ideologie und Utopie*. Bonn: F. Cohen, 1929.)
13. Merton, Robert K. "Karl Mannheim and the Sociology of Knowledge." In Robert K. Merton, *Social Theory and Social Structure*. Glencoe, Ill.: Free Press, 1957. pp. 489–508. (Originally published in 1941.)
14. Merton, Robert K. *Science, Technology, and Society in Seventeenth Century England*. New York: Howard Fertig, 1970. (Originally published in 1938.)
15. Merton, Robert K. *The Sociology of Science*. Chicago, Ill.: University of Chicago Press, 1973.
16. Shapin, S., and A. Thackray. "Prosopography as a Research Tool in the History of Science: The British Scientific Community, 1700–1900." *History of Science* 12 (1974): 1–28.

# 16

# Scientists and the Social Study of Science:
# A Research Problem

I start with a personal problem: during some forty years of interest and work in the social study of science I have often felt dismay and even anger at the widespread ignorance among natural scientists about this field, their seeming denial that in principle it is even possible, their arrogant assertion of all kinds of sociological, political, and psychological generalizations about science without any awareness of the limitations of their impressions, their prejudices, and their commonsense knowledge.

Let me try today to turn my personal problem into a sociological problem. Can we move toward defining some research in which not only the beliefs and attitudes of scientists toward the social study of science are systematically determined but also some of the social and cultural sources of those attitudes? I think we can, and as a step along the way, let me offer some impressionistic evidence and some preliminary hypotheses about why there is so much ignorance, negativism, and ambivalence among natural scientists toward the social study of science. Survey research, intensive interviews, and documentary sources could all be used for collecting data.

My impressionistic evidence is not representative of the views of all scientists. I have taken materials from scientists who might be called "the best and the brightest" of their kind, men who are perhaps of special concern for us just because of their articulate and forceful ideologies on this matter. I start with C. P. Snow,[1] whose views started a universal and heated debate. "I believe," said Snow, "the intellectual life of the whole of western society is increasingly being split into two polar groups. . . . Literary intellectuals at one pole—at the other, scientists, and as the most representative, the physical scientists." By leaving out the social scientists altogether, Snow nullified the possibility that the social study not only of science but of other social phenomena might exist.

From Snow we move to more recent times and to a more ambivalent scientist, John Ziman, a distinguished physicist and scientific administrator.[2] Ziman is intimately familiar with a considerable amount of the good work in the history, philosophy, and sociology of science. He seems to accept a good deal of it. Yet there is a deep ambivalence in his final views, if we may judge from his conclusions on the last four pages of his book:

> In this search for "reliable knowledge" concerning human behavior we thus arrive at a rather disappointing conclusion. There are grave obstacles of categorical vagueness . . . , operational variation . . . , experimental irrelevance . . . , theoretical unprovability . . . and cultural relativism in the way of establishing a general "science of society." . . . This is not to deny the value of sociological and anthropological research, which uncovers such richness and variety, such poverty and uniformity, in the lives of all human beings. In default of such research, we should live entirely at the mercy of our own follies, misconceptions and deceptions as to the merest facts of the matter.[3]

> . . . Social commonsense is a very satisfactory guide to social action: "Everyone is, in a certain sense, a fairly competent social scientist. . . ." [This last is a citation from R. Harre, *Reconstructing Social Psychology*.][4]

> . . . What we live by in social life is not a series of scientific laws, but *maxims*, whose lack of consistency and consensuality is irrelevant to their practical values.[5]

> . . . The novelist, with his sensible ear and discriminating eye, articulates the universal elements in our emotional lives, and teaches us more about mankind than any formal theory.[6]

> . . . we cannot learn the art of research by reference to the formal philosophies, sociologies and psychologies of science. . . . We read our Popper and our Kuhn, our Merton and our Polanyi, not for rules and laws and formulae and proofs, but for maxims and insights and understanding. We also turn to the makers themselves—to Newton and Faraday, to Poincaré and Einstein, to Darwin and Pasteur, to Freud and Jung, and Durkheim and Piaget, to rediscover the significance of their contributions to the world they have helped create.[7]

This is at best a deep ambivalence, at worst a forbidding negativism about the worth of theoretically derived and empirically tested social study of science.

Freeman Dyson is another distinguished and eloquent physicist. His book *Disturbing the Universe*[8] is the first of a series sponsored by the Sloan Foundation as part of its long-standing program to encourage "the public understanding of science." Although the "Preface to the Series" says that "an understanding of the scientific enterprise, as distinct from the data and concepts and theories of science itself, is certainly within the grasp of us all," it gives not a single reference to any social-science expertise in the field. As for Dyson himself, while he seems to admit the existence of the social sciences in the following statement, he himself will not have any of them. He says: "My colleagues in the social sciences talk a great deal about methodology. I prefer to call it style. The methodology of this book is literary rather than analytical. For insight into human affairs, I turn to stories and poems rather than to sociology." And, continuing, Dyson rightly recognizes the sources of his preference. "This is the result of my upbringing and background. I am not able to make use of the wisdom of the sociologists because I do not speak their language." But is this language something a man like Dyson couldn't master with not the greatest of efforts? It is, rather, that his heavily classical and literary education in England has always seemed sufficient to him and he felt no compelling reason to take the small effort to read in the works of the social scientists of science.

Part of Dyson's preference may be due to the strong individualistic cast of his view of the world. "To understand the nature of science and its interaction with society," he says, "one must examine the individual scientist and how he confronts the world around him. . . . my individualistic bias leads me to listen to poets more than to economists." We see that Dyson is himself aware, as many scientists are not, of his individualistic theory about social institutions and social action. Another assumption of Dyson's that may contribute to his uneasiness with the social study of science is his unpredictability assumption that "Science and technology, like all original creations of the human spirit, are unpredictable." This, of course, is an assumption that the social study of science cannot accept.

Looking at one last distinguished scientist, P. B. Medaware,[9] we find him not only aware of social science but willing to admit that it might improve upon his "mere" opinion. He says, "I use the word *opinion* to make it clear that my judgments are not validated by systematic sociological research and are not hypotheses that have already stood up to critical assaults. They are merely personal judgments, though I hope that some of them will be picked up by sociologists of science for proper investigation." We may hope that Medawar represents the wave of the future among scientists in their view of the social study of science.

Now let us look at a series of preliminary hypotheses about why it is that we see so much negativism among scientists about the social study of science. The factor of their socialization and early training, mentioned by Dyson, suggests one hypothesis to be tested by research. This is that scientists are likely, if they are trained at all outside of science itself, to have studied the humanities rather than the social sciences, and especially not the social studies of science, which are still not taught widely or well at most colleges and universities. Very few scientists have had any work as undergraduates in social science, probably not even economics. And there has been no encouragement from their mentors to study social science at some more advanced level.

Research should investigate whether generational differences are occurring. Are younger scientists more receptive to the social study of science?

Second, we may offer the hypothesis that the strong values of individualism and autonomy in science lead to resistance to the sociological premise of a considerable measure of social determinism in the creation of scientific knowledge, in the scientific reward system, and in the social organization of science. Do the social explanations of science seem to undermine the values and committment of scientists as they go about their everyday tasks? This does not seem to happen to the social scientists of science as they go about their researches.

Third, is it a correct hypothesis that all powerful groups tend to resist objective scrutiny by outsiders? Are we all aware that there are dangers in exposing the fact that our values are not fully realized, that much of our performance is routine, that there is even some dishonored deviance? The actual working of institutions and roles is always somewhat different, for specifiable social reasons, from what the official public values and statements of the institutions hold it to be. All practicing insiders know about such things, but public visibility is seen as dangerous to the prestige and the interests of the group. Is this one of the reasons why natural scientists are averse to the objective scrutiny of the social scientists of science?

Fourth, do the views of natural scientists on the social study of science come from an inadequate understanding of science? Do scientists hold the simplistic view that "a scientist" can understand anything just because he is "a scientist"? Do they feel that he needs no special knowledge that he can't work up himself and especially for social and political problems? Do natural scientists understand the functions of different technical conceptual schemes for different aspects of reality?

As these hypotheses and others are explored and tested, it will be well to explore exceptions to the patterns of ignorance, indifference, and ambivalence I have reported. "Deviant case analysis" can be used in place of controlled experiment, as "natural experiment," as a basis for comparative proof of the social sources of the predominant patterns. How, for example, did Thomas Kuhn come to recognize explicitly that his theory of scientific change and revolution was sociological as well as philosophical and historical? How did David Edge, who was a practicing radio astronomer, come to look favorably on the social study of science and come to collaborate with Michael Mulkay?

The future progress of the social studies of science will be easier, I think, if there can be more cooperation between natural and social scientists. As we try to understand how to promote such cooperation, we may be able to learn from past failures the conditions under which cooperation does not and does occur. For that reason, research into the attitudes of scientists toward the social study of science looks attractive for its own sake and as a guide to more effective cooperation.

## Notes

1. C. P. Snow, *The Two Cultures and the Scientific Revolution* (Cambridge, England: Cambridge University Press, 1959).
2. John Ziman, *Reliable Knowledge: An Exploration of the Grounds for Belief in Science* (Cambridge, England: Cambridge University Press, 1978).
3. Ibid., pp. 183–84.
4. Ibid., p. 184.
5. Ibid., p. 185.
6. Ibid.
7. Ibid.
8. Freeman Dyson, *Disturbing the Universe* (New York: Harper & Row, 1979).
9. P. B. Medawar, *Advice to a Young Scientist* (New York: Harper & Row, 1979).

# 17

# Tension and Accommodations between Science and Humanism

Ever since the seventeenth century, the relative emphasis on science and humanism in the educational curriculum has been a problem, and often a central one, for Western institutions of higher learning.[1] The tension and conflict between science and humanism have, of course, been often noted and endlessly discussed, but still seem much less well understood than they might be.

Two types of erroneous views are recurrently advanced and defended with some passion. One is that there is no essential difference between science and humanism, that humanists and scientists are pretty much the same kind of social animal, and that, therefore, there is no reason at all, except ignorance or ill-will, for tension between them. René Dubos, for example, recently wrote: "As far as I can judge, science meets all the requirements usually associated with the concepts of culture and humanism."[2]

The second erroneous view is that the differences between science and humanism are so great and so fundamental that only the crippling, or even the total destruction, of the other will enable the one to carry on its activities. In recent times, this view is more likely to be held by a humanist, though the case was just the opposite about a hundred years ago, when science seemed the weaker party in the conflict.

A proper sociological view of the perennial tension between science and humanism requires a functional analysis of the essential place of each in society. By defining them in terms of functional characteristics, we are able to specify both their relative autonomy and their necessary interdependence.[3] It is only by understanding the culturally and socially patterned sources of any particular kind of social tension that we can best construct appropriate forms of accommodation for the conflict which issues from it.

In addition to our functional analysis, some historical perspective on the problem will be valuable. The quarrel between science and humanism is an ancient and honorable one. Its origins in the seventeenth century, if not earlier, and its persistence through a variety of historical periods, though with varied intensity and form, add credibility to the suggestion that the cultural and social sources of the tension are constant.[4] Historical illustration is not being used here for its own sake, but for comparative purposes, to bring out not only the continuity of the tension but some of its repeated foci in otherwise quite different historical conditions.[5] Through comparative historical materials, then, we seek some of the defining characteristics of science and humanism, characteristics which become, in part, structural sources of tension between the two. Before discussing four polar sets of these characteristics in detail, some limitations of our analysis should be made explicit. These qualifications would be necessary in any context; they are all the more important in this one because of the fact that our key phenomena, "science" and, even more so, "humanism" have been defined in so many different ways by so many of those who have discussed the tension between them.

First, our four polar sets of characteristics are not intended to be exhaustive of all, nor even the most important, cultural and social characteristics of science and humanism. And, second, our characteristics are intended to be analytic and may not correspond precisely to any currently established, concrete categories of individuals who call themselves or others "scientists" and "humanists". Thus, as will become clear, our definition of "science" probably will include all natural scentists but will very likely exclude some of those who call themselves "social scientists." Sociologists, for example, fall on both sides of all four of our polar sets of characteristics. In our terms, some sociologists would be considered "scientists," others "humanists."[6] Many scholars—but probably a minority—in such fields of specialization as English literature, linguistics, and history, are "scientists" by our definition. In some fields of specialization in the humanities, and the field of history is as good an example as any, tension between the "scientists" and the "humanists," as we would define them, is one of the central problems within the discipline as well as for its relations with the world outside.[7] Indeed, the tension between "science" and "humanism" may show itself throughout the work of a single individual.[8]

## Cultural Characteristics

Science and humanism, then, may be defined in terms of three sets of polar pairs of *cultural* characteristics. The first set is: *direct concern with values*, for humanism, versus *a morality*, for science. The second set is: *concreteness*, for humanism, versus *abstractness*, for science. And the third set is:

*pessimism*, for humanism, versus *optimism*, for science. These sets seem, further, to be related to one another. Thus, there is a tendency for concern with values, concreteness, and pessimism to go together in humanism; and a similar tendency for a-morality, abstractness, and optimism to cluster in a syndrome of defining characteristics for science. As we analyze each of these sets in detail, we shall also analyze the connections between them.

*Direct Concern with Values versus A-morality*

The distinction we are making here has its predecessors in distinctions such as Cassirer's between "thing-perception" and "value-perception"; or John Dewey's between "scientific thinking" and "qualitative thinking"; or, finally, by Talcott Parsons, among "instrumental orientations," "value orientations," and "expressive orientations."

On the functional view, values are essential in society as patterned principles for making those choices that inevitably confront goal-seeking men as they carry out the activities that are necessary to create and keep their social structure and culture in being. As values are functional, so also are those roles and their occupants who specialize, more or less explicitly, in making clear what is the pattern of values in a society and what are its applications to concrete situations of choice. In sum, there has to be some specialization in direct concern with values in any society, and this is what we are defining as the "humanist" concern. Any threat, apparent or real, to deny the functional necessity for a direct concern with values is seen as threatenting by humanists.

In the seventeenth century, for example, when Bishop Sprat's *History of the Royal Society* and Joseph Glanvill's *Plus Ultra* were propagandizing for the "new philosophy" and, seemingly, denying the difference between the scientific and moral worlds, the humanist Meric Casaubon was only one intelligent critic of "the belief that experimental science could 'moralize' men." In the eighteenth century, Dr. Johnson put the humanist view quite succinctly in his famous epigram: "We are moralists by nature, geometricians by chance."[9]

Contemporary statements of the humanists' view on values abound. E. A. Burtt, for example, has said: "The mediaeval mind was dead right in its conviction that science is not an independent enterprise, free to follow its lone course irrespective of how it affects other human values. . . . Science is not free to pursue such an outcome, in blithe disregard of the other moral and social goods that lure men onward and claim their allegiance."[10] A less angry and more positive humanist view has been given in his excellent little book by Moody Prior: " . . . the humanities address themselves to an understanding and an evaluation of human goals. . . . By comparison, a scientific generalization . . . carries no implication within itself of its relevance to any human uses to which it may be put, to the human choices which may be governed by it, or to the inherent human striving for happiness or self-fulfillment in action. . . . The creations of science . . . are neutral with reference to their moral and social implications."[11]

For just this reason, among contemporary sciences, or aspirants thereto, the social sciences have become the *bêtes noires* of some humanists, who see these social sciences as denying the existence and importance of human values.[12] And those social scientists whose work shows a direct and explicit concern for values, men like C. W. Mills, David Riesman, and W. H. Whyte, Jr., are admired by the contemporary humanists who disdain more "scientific" social scientists.[13]

But if a direct concern with values has its autonomous functions in society, a-morality does so also, in certain special social contexts. Science holds that values are not in any direct way relevant to the cognitive understanding and control of the empirical physical, biological, and social aspects of the world. "While commonsense knowledge is largely concerned with the impact of events upon matters of special value to men," says Ernest Nagel, "theoretical science is in general not so provincial. The quest for systematic explanations requires that inquiry be directed to the relations of dependence between things irrespective of their bearing upon human values. . . . Theoretical science deliberately neglects the immediate values of things, so that the statements of science often appear to be only tenuously relevant to the familiar events and qualities of daily life."[14] And Charles Gillispie has said: "For neither in public nor in private life can science establish an ethic. It tells what we can do, never what we should. Its absolute incompetence in the realm of values is a necessary consequence of the objective posture."[15]

Just as humanism has seen itself struggling against science to maintain the functional autonomy of its direct concern with values, so science has seen itself as struggling against humanism to establish cognitive maps of the world free from the taint of values. Alexander Koyré has said that the seventeenth-century revolution in science effected "the discarding by scientific thought of all considerations based upon value-concepts, such as perfection, harmony, meaning, and aim, and finally the utter devalorization of being, the divorce of the world of value and the world of facts."[16] But this conclusion—that values have been entirely excluded from where only the a-morality of science should prevail—is premature, if we may accept Gillispie's historical argument in *The Edge of Objectivity*. First the physical sciences, then the biological, and now the social have been struggling against what they consider the humanist intrusion of values into "the order of nature" all during the last three hundred years. To use Gillispie's metaphor, it has been hard to advance "the edge of objectivity" into those value-filled physical, biological, and social worlds that men in modern society have accepted as firmly, or nearly so, as their ancestors.

Some scientists, to be sure, have felt that the struggle was ended, that all values had been entirely driven out of the empirical world. A variety of "positivistic" views of the world have held that science alone is adequate for man's adaptation to his situation. The chemist Marcelin Berthelot, for example, claimed in 1901 that: "Science is today in a position to claim the leadership of societies, not only with regard to material questions, but also to intellectual and moral problems. . . . It is science that will provide the truly human basis of morals and politics in the future."[17] Such extreme views are now held by few scientists, and very seldom with such conviction.

The functional tension that exists between science and humanism with regard to values is inevitable. Because each has its special functions, there are limits on what each of them can responsibly claim as its special sphere. Many scientists have been unwilling or unable to cope with these limits. It has been suggested that the appeal to scientists of Baconianism, that is, of humanitarianism in science, is owed to their unwillingness to face the limits which the nature of science places upon scientists as scientists.[18] Of course, the scientist is, as a concrete individual, also a member of some society, a "citizen" as we now say, and in this other role or set of roles he has a direct concern with values. But these roles are, from the point of view of his scientist role, secondary concerns. And in any case, the two sets of concerns are not to be confused one with another. Special competence in the scientist role, with its a-morality, does not necessarily and alone give special competence in the "citizen" or "humanist" role, where values are essential.[19]

Scientific knowledge is, of course, always functionally interdependent with values. Better scientific knowledge is better for better value choices. Science, natural and social alike, is endlessly creating new knowledge for which there must be a restatement of values by humanists. This is the proper function of humanism. But when humanists have been making value choices without the best available scientific knowledge, scientists have become impatient and rushed in with their own value choices. Yet these choices might be just as inadequate, if they were as defective with regard to values as the humanist choices were with regard to scientific knowledge. The obvious pattern of accommodation here is one in which different types of functional specialists collaborate to produce the functionally best synthesis. Such a pattern will always be difficult to achieve in any particular situation, but in general it will be helpful if scientists and humanists at least know what the functional constituents of the pattern have to be.

*Concreteness versus Abstractness*

Whereas humanism tends toward concreteness in the structure of its thought, science strives for as great abstractness as possible. The functions of abstractness for science have been put very clearly by Ernest Nagel:

> . . . the unusually abstract character of scientific notions, as well as their alleged "remoteness" from the traits of things found in customary experience, are inevitable concomitants of the quest for systematic and comprehensive explanations. Such explanations can be constructed only if the familiar qualities and relations of things, in terms of which individual objects and events are usually identified and differentiated, can be shown to depend for their occurrence on the presence of certain other relational or structural properties that characterize in various ways an extensive class of objects and processes. Accordingly, to achieve generality of explanation for qualitatively diverse things, those structural properties must be formulated without reference to, and in abstraction from, the individualizing qualities and relations of familiar experience.[20]

It is of the essence of science, thus, to be systematic, and to be systematic requires it to be highly abstract and see similar properties in what are apparently diverse objects. This abstractness ignores the familiar and the moral qualities of objects, which are so important to humanists.

A somewhat less general, but no less trenchant view of the scientific disposition to abstractness has been stated by a social scientist, Robert K. Merton. "In no sphere of systematic knowledge," says Merton, "whether it be mechanics, biology, linguistics, or sociology, do specialists go on the fool's errand of explaining every aspect of concrete phenomena. Instead, particular aspects, structures, and processes of the phenomena are singled out, under the guidance of some general ideas, and methodically investigated, while other aspects are conscientiously neglected as no part of the problem in hand."[21]

In contrast to the abstractness of science, the autonomous function of humanism is to strive for concreteness. "In science," says Prior, "the total absorption of the individual event in the generalization is the goal; on the other hand, the humanities are concerned rather with providing for the special meaning of the indivdual event within an appropriate general system."[22] To understand the special moral meaning of events in the concrete world requires the application of several generalizations in a concrete pattern of evaluation. No one abstraction will serve if the humanist function of moral evaluation is to be properly carried out. Humanism thus needs the abstractions of science, but it always applies them concretely to the world it is evaluating. Whereas the function of science is to be interested in abstractions for their own sake, it is the function of humanism to be interested in them only as they can be applied to the making of moral choices in the world of concrete and individual

events. The humanist preference for concreteness is expressed in its preference for "totalities" and "wholeness." Moral problems come as concrete
wholes, not as abstractions.

Humanists often have not understood this tension between abstractness and
concreteness. Certainly it must have been a humanist who first coined the
phrase, "bloodless abstractions." And only recently a humanist spoke of
those "two irreconcilable worlds—the world of intimate experience and that
world of abstract convictions in which the validity of intimate experience is
categorically denied."[23] But the abstract and the concrete aspects of the physical, biological, and social worlds are not irreconcilable with one another.
There is a pattern of accommodation which recognizes the functional autonomy of each and, equally, their necessary interdependence.

Even when they have recognized the general tension between the abstractness of science and the concreteness of humanism, the humanists have sometimes lamented the destruction of some of their particular and established
concretenesses by the new abstractions of science. The following lines from
Donne's *First Anniversary* (1611), especially the last four, express for all time
the dismay of the humanist against the crumbling of the established world by
new abstractions:

> And new Philosophy calls all in doubt;
> The Element of fire is quite put out:
> The Sun is lost, and th' Earth, and no man's wit
> Can well direct him where to looke for it.
> And freely men confesse that this world's spent,
> When in the Planets, and the Firmament
> They seek so many new; they see that this
> Is crumbled out againe to his Atomies.
> Tis all in peeces, all cohaerence gone;
> All just supply, and all Relation.

The endless humanistic complaints about jargon in science can also be seen
as an expression of the tension between concreteness and abstractness as
functional patterns of thought. Humanists attach great importance to the established concrete language as the repository of the received morality and the
experienced beauty of their time. But scientists, on principle, distrust the
established language *because* it embodies and sanctifies the received ideas
which they are trying to analyze into new abstractions. Insofar as jargon
consists of genuinely new abstract ideas, it is essential to science. "No science can flourish," says Gillispie, "until it has its own language in which
words denote things or conditions and not qualities, all loaded with vague
residues of human experience."[24] A science is the more scientific the more
nearly its abstract terminology is incomprehensible to common sense. The

necessity for accomodative cooperation between science and humanism in translating their different languages one into another is obvious.

*Pessimism versus Optimism*

Finally, whereas science tends toward optimism in its view of its tasks in the world, humanism tends toward pessimism so far as its own special functions are concerned. Science, we have seen, consists in the construction of ever more generalized, more systematic, and more nearly exhaustive conceptual schemes for describing the empirical world. Since in principle it is possible to achieve these more generalized, systematic, and nearly exhaustive structures of thought, science is in principle progressive and evolutionary. In the case of science, optimism, or the view that one can master the special tasks that are set by one's activity, breeds on what it produces. The history of science has in fact seen the realization of a continuing evolutionary progress in scientific ideas and an accompanying enlargement of scientific optimism. The fullness of knowledge, Bacon said, lies in the fullness of time. Scientists feel that their experience has verified this aphorism; they feel as optimistic about the present and future as they feel confident about the past.

Only a man with scientific optimism, never a humanist, would have written as Henry Power did back in the seventeenth century:

> You, most Noble Souls, the true Lovers of Free and Experimental Philosophy, you are the enlarged and Elastical Souls of the World, who . . . do make way for the Springy Intellect to flye out into its desired Expansion. This is the Age in which all mens Souls are in a kind of fermentation, and the spirit of Wisdom and Learning begins to mount and free it self from those drossie and terrene Impediments wherewith it hath been so long clogg'd. These are the days that must lay a new Foundation of a more magnificent philosophy.[25]

Robert Wood in his analysis of scientists in politics attests that this optimism still exists today: "Yet, though since Hiroshima scientists have 'known sin,' there is little evidence that their faith has been shaken in the essentially benign effects of expanding knowledge or the ennobling character of their calling."[26]

Humanism, in contrast, is concerned with ideas about values, with the difficult choices values entail, and with the inevitable limitations on their fulfillment. What theologians call the "problem of meaning" is inherent in human life; so is the sense of human finitude; and these and other moral problems are what are always there for humanism. When the humanist speaks of "the human condition," he means the tragic elements of existence, the evils that cannot be overcome, the goods that cannot be achieved, in sum, the persistence of value problems. If there has been any evolutionary progress in human value problems, few have claimed to see it, and certainly not the humanists. Here again, as with science, the functional task and its social products breed the view that prevails among the practitioners. But in this case what prevails is pessimism, or the view that only with great difficulty can one even cope with the tasks set by one's activity. Humanists see a fundamental sameness about the problems of the human condition, past, present, and future, a sameness that includes evil, tragedy, limitation, and the endless recurrency of value problems.

## Science as a Social Movement

Our fourth and final polar, differentiating characteristic between science and humanism is a social one. For the last three or four hundred years at least, science has tended to be an exuberant, optimistic, triumphant social movement, seemingly sweeping all before it in its success, not only replacing other ideas with its new abstractions but capturing ever more social positions of influence, for example, in the universities and in government. Humanism, on the other hand, has tended to be divided, defensive, with intellectual and social resources scattered, ill-organized, and conservative.

Systems of ideas and social systems alike undergo processes of differentiation and specialization. The great advances of science during the last three hundred years, we can now see, have resulted in considerable measure from just such processes of differentiation and specialization out of humanist ideas and humanist roles. First our scientific ideas about the physical world were successfully differentiated out from moral and religious concerns; then our ideas about the biological world; and now, finally, our ideas about the social world. As for the role of scientist, its occupants hardly existed in the seventeenth century but have grown in number, ever since, in what is apparently a geometrical ratio.[27] The innovations produced by these processes of differentiation and specialization would probably, in any case, have caused tension and resistance among those who saw older ideas overthrown, older social positions of influence weakened or destroyed. Some tension and resistance are probably inevitable as a result of processes of social and intellectual differentiation.[28] Tension resulting from differentiation processes may result in conflict, but it may also be resolved by various patterns of accommodation in which the disturbing effects of innovation are the more effectively cushioned because its sources are understood.

Thus science has been more than a development of ideas differentiating out of humanism. It has been, in addition, a social movement triumphantly carrying these ideas forward into new contexts and new social roles. It has been a social movement in somewhat the same way that a variety of enthusiastic religious and ideological movements have been, winning new scope and influence for their new ideas. Already in the seventeenth century, science as an enthusiastic social movement was gathering strength. "Many biographies relate the feeling of confidence," says Hall, "the depth of intellectual satisfaction, the release of a creative drive that was experienced by members of the new class of layman, educated and leisured, when they passed from the confines of academic disputation to the methods of experimental science."[29] And Lynn Thorndike reports: "A remarkable feature of both scientific and pseudo-scientific works of the seventeenth century is the frequency with which such words as 'new' and 'unheard-of' appear in their titles."[30] In England at this time, science as a social movement centered on the Royal Society; Sprat's *History of the Royal Society* and Glanvill's *Plus Ultra* were self-conscious and ardent pieces of propaganda for the Royal Society and for science, asserting the superiority of the "new philosophy" to the ancient and established ideas of the humanists.[31] We have already seen how Meric Casaubon, among other humanists, tried to respond to these upstart and self-praising "moderns," as the scientists or "new philosophers" were glad to pose themselves in oppostion to the inferior, dull, and resistant "ancients" or humanists.

The scientists retain their exuberance in the twentieth century, as their ideas continually develop and as their social movement wins ever new influence in the universities and in government. They have the future in their bones, says Snow, speaking now for his fellow scientists. He reports that "archetypal figure, Rutherford, trumpeting: "This is the heroic age of science! This the Elizabethan age!"" Or again, he tells that when somebody said: "Lucky fellow, Rutherford, always on the crest of the wave," Rutherford triumphed with, "Well, I made the wave, didn't I?"[32] Some humanists try hard to tolerate such exuberance. "Scientists cannot help being impressed with the importance of their work and the fabulous character of their activities," says Prior, "and it is their obligation to seek the advancement of science."[33] But other humanists are often defensive, frightened, and angry in the face of a social movement which continually weakens some of their ideas and seems to threaten to usurp their social position altogether.

## Conclusions

Science and humanism as we have defined them in terms of four polar pairs of cultural and social characteristics, each has its own autonomous functions to perform in society. Inevitably, however, as they carry out their essential tasks they also partially limit one another. This creates tension, a tension which is structured by the very nature of what is at once the independence yet interrelationship of science and humanism. The amount and strength of the tension varies. It is likely to be greatest and strongest when a whole new science or set of sciences is undergoing a revolution, that is, when a whole new set of abstractions about some aspect of the empirical world is being created. So it was in the seventeenth century, for physical science. So it seems to be now in the twentieth century, for the several social sciences. Whether or not tension will be as great as it has been on these historical occasions, some tension must always be present. And since this is so, since there is a structured strain between science and humanism, both must learn to live with what cannot be eliminated. Neither disregard of the tension, nor utopianism about the chances for eliminating it, nor unlimited hostility on either side will serve to construct suitable patterns of accommodation between the two, which is the best that can be achieved. A challenging intellectual and moral task for our time, as for times past, is to make manifest these somewhat latent sources of tension and to seek continuously for the compromises, the understandings, in short, the patterns of accommodation between science and humanism.

## Reference

A version of this paper was presented at a meeting sponsored jointly by Section K, American Association for the Advancement of Science, and the American Sociological Association, held in Philadelphia, Pennsylvania, on December 29, 1962.

## Notes

1. For the seventeenth century, see Richard Foster Jones, *Ancients and Moderns: A Study of the Scientific Movement in Seventeenth-Century England*, 2nd ed. (St. Louis: Washington University Studies, 1961), esp. chap.9, The "Bacon-Faced Generation"; for the eighteenth, see Nicholas Hans, *New Trends in Education in the Eighteenth Century* (London: Routledge & Kegan Paul, 1951); and, for the nineteenth, Alan Price, *Humanities versus Science in Mid-Nineteenth Century Educational Thought in England*, (unpublished Ph.D. dissertation, Queen's University of Belfast, 1957), and Cyril Bibby, *T. H. Huxley: Scientist, Humanist, and Educator*, New York, Horizon, 1959.
2. Dubos, *The Dreams of Reason*, (New York: Columbia University Press, 1961) p. 160.
3. Note the similarity of this approach to that stated by Dean Moody Prior of Northwestern University, who is a humanist Professor of English: "There are important functions which the humanities cannot perform, and there are important functions which science cannot perform . . . ," in *Science and the Humanities* (Evanston: Northwestern University Press, 1962), p. 22.
4. For a brief comment on still earlier tension between science and humanism, see the statements about Plato and Aristotle in C. C. Gillispie, *The Edge of Objectivity* (Princeton: Princeton University Press, 1960), pp. 13–14.
5. Even C. P. Snow, who prides himself on bridging what he calls "the two cultures," seems to be unaware of the fact that what is so pressing a problem to him now has a long historical antecedence. Moody Prior, however, *op. cit.*, p. v.,

speaks of "this ancient question."

6. For a passionately "humanist" declaration by a sociologist, see Ralf Dehrendorf, "Democracy without Liberty: An Essay on the Politics of Other-Directed Man," in S. M. Lipset and Leo Lowenthal, eds. *Culture and Social Character: The Work of David Riesman* (New York: Free Press of Glencoe, 1961). "Scientific" statements are, on the other hand, legion among sociologists. See, for example the works of Talcott Parsons, Robert K. Merton, or Paul F. Lazarsfeld.

7. A typical and important recent statement of this problem which comes down firmly on what we would call the "humanist" and not the "scientist" side in history can be found in the foreword by Dexter Perkins to *The Education of Historians in the United States* by Perkins and John L. Snell, (New York, Mc-Graw-Hill, 1962).

8. When it does, it is an example of one of the several kinds of sociological ambivalence which has been so instructively analyzed in Robert K. Merton and Elinor G. Barber, "Sociological Ambivalence," in Edward A. Tiryakian, ed., *Essays in Honor of P. A. Sorokin* (New York: Free Press of Glencoe, 1963).

9. See Jones *Ancients and Moderns*, pp. 241ff.

10. Burtt, "The Value Presuppositions of Science," *Bulletin of Atomic Scientists* 13 (1957): 99–106, reprinted in P. C. Obler and H. A. Estrin, eds., *The New Scientist* (New York: Anchor Books, 1962).

11. Prior, *Science and the Humanities*, p. 18; see also, pp. 11, 16, 58. For some other influential statements by humanists, see Douglas Bush, *Science and English Poetry* (New York, Oxford University Press, pp. 5, 165; and Richard Schlatter, "The Ford Humanities Project at Princeton: The Job of the Humanist Scholar," in *ACLS Newletter*, 12, no. 10 (1961): 6.

12. See Joseph Wood Krutch, *The Measure of Man: On Freedom, Human Values, Survival, and the Modern Temper* (Indianapolis: Bobbs-Merrill, 1953), pp. 31, 171, 191 et passim.

13. On the social sciences and the humanities, further, see Bernard Barber, *Science and the Social Order*, rev. ed. (New York: Collier Books, 1962), pp. 337ff.

14. Nagel, *The Structure of Science* (New York: Harcourt, Brace & World, 1961), pp. 10–11.

15. Gillispie, *The Edge of Objectivity* p. 154.

16. Koyré *From the Closed World to the Infinite Universe* (Baltimore: Johns Hopkins University Press, 1957), p. 2

17. Quoted in Dubos, *The Dream of Reason*, p. 9.

18. Gillispie, *The Edge of Objectivity,* pp. 81–82

19. For a discussion of some of these confusions and their consequences in the political realm in recent times, see Robert Gilpin, *American Scientists & Nuclear Weapons Policy* (Princeton: Princeton University Press, 1962).

20. Nagel, *The Structure of Science*, p. 11.

21. Robert K. Merton and Robert A. Nisbet, eds., *Contemporary Social Problems*, (New York: Harcourt, Brace & World, 1961), p. 700. On the social-science view of abstraction, further, see Robert E. Lane, *The Liberties of Wit: Humanism, Criticism, and the Civic Mind* (New Haven: Yale University Press, 1961), esp. pp. 128–29.

22. Prior, p. 15; see also Schlatter, *Ford Humanities*, pp. 5, 8, 9.

23. Krutch, *Science*, p. 194.

24. Gillispie, *The Edge of Objectivity*, p. 77.

25. See the conclusion to Power's *Experimental Philosophy*, 1664, cited in Marjorie

Nicolson, *The Breaking of the Circle* (New York: Columbia University Press, 1960), p. 203. For a statement of the pessimistic views of seventeenth-century humanists, see Douglas Bush, "Science and Literature," in H. H. Rhys, ed., *Seventeenth Century Science and the Arts* (Princeton: Princeton University Press, 1961), p. 34.

26. Wood, "Scientists and Politics: The Rise of an Apolitical Elite," unpublished paper, 1962. On the scientists' optimism, see also C. P. Snow, *The Two Cultures and the Scientific Revolution*, London: Cambridge University Press, 1959.

27. See Derek Price, *Science since Babylon*, (New Haven: Yale University Press, 1961), chap. 5, "Diseases of Science," for a discussion of the rate of growth of scientific roles and ideas.

28. Resistance occurs not only between scientists and humanists, but also among scientists themselves when intellectual innovations occur. See Bernard Barber, "Resistance by Scientists to Scientific Discovery," *Science* 134 (1961): 596–602.

29. A. R. Hall, *The Scientific Revolution, 1500–1800* (London: Longmans, Green, 1954), p. 187.

30. "Newness and novelty in seventeenth-century science and medicine," in P. P. Wiener and Aaron Noland, eds., *Roots of Scientific Thought* (New York: Basic Books, 1957).

31. See Jones, *Ancients and Moderns*, pp. 183ff., 241ff.

32. Snow, *The Two Cultures*, pp. 4–5.

33. Prior, *Science and the Humanities*, p. xii.

# Index